COMMON MISCONCEPTIONS IN ELECTRODYNAMICS

Juan R. González Álvarez

COMMON MISCONCEPTIONS IN ELECTRODYNAMICS

Edition: First hardcover (February 2024)
Series: Common misconceptions in physics
Imprint: Independently published
ISBN: 9798321091333

DISCLAIMER

This book was carefully produced. Nevertheless, the author does not warrant that the information contained therein is free of errors. Readers are urged to be aware that statements, data, illustrations, procedural details or other items may be inadvertently inaccurate.

ALSO BY THE AUTHOR

Common misconceptions in **special relativity**
thermodynamics
general relativity
statistical mechanics
quantum mechanics
quantum field theory

Review of **the God equation**
ten keys to reality

ABOUT THE AUTHOR

Juan R. González Álvarez studied physics and chemistry at the *Universidade de Vigo*. He worked in scientific organizations such as the *Ilustre Colegio de Químicos de Galicia* and was a research assistant at the *Instituto de Investigaciones Marinas (IIM-CSIC)* in the biogeochemistry and hydrodynamics of estuaries, participating in several congresses, reports, and monographs. Juan has also conducted research on the nature of heat, has developed his own formulation of quantum mechanics, an alternative to the dark matter model, has also closely followed the development of string theory during the last three decades, and is now developing a multiscale dynamics for the unified description of physical, chemical, and biological processes. Juan and his family live in Galicia, in the city of Vigo.

Website: https://juanrga.com
X/Twitter: @juanrga

ACKNOWLEDGMENTS

I would like to thank Professor J. M. van Ruitenbeek and to Professor John Roche for reviewing parts of this book. Roche's incisive suggestions were also useful in the rest of the book. I also thanks to Professor Robert M. Wald for clarifying the interpretation of Lagrangians and Hamiltonians in his book, which helped to improve chapter 4.

I would like to express my deep gratitude to John Stout for the many helpful comments and suggestions that helped me greatly improve the presentation and the content of the whole book.

Any remaining errors are my sole responsibility.

CONTENTS

LIST OF FIGURES AND TABLES

PREFACE

When you first learn physics —for example in school— you have many questions, but you expect them to be answered in more advanced treatises. Ironically, more advanced presentations of physics raise many more questions. You could say that you know more, but not that you known better.

Furthermore, both introductory and advanced physics treatises tend to avoid mentioning those problematic aspects that invalidate an idealized vision of this discipline. Virtually everyone interested in physics knows about the incompatibility of general relativity and quantum mechanics, but have you ever heard of the incompatibility of general relativity and thermodynamics or the incompatibility between quantum mechanics and classical mechanics? More than one hundred and thirty years have passed since Lorentz published in "*La théorie electromagnétique de Maxwell et son application aux corps mouvants*" the force that bears his name –the Lorentz force– and, even today, experts in electromagnetism do not agree on the correct equation of motion for a charge subject to the action of an electromagnetic field. I have been told many times that physicists are very conservative, but that is not an excuse to hide from students and the general public the main problems and deficiencies of physics.

The primary goal of this book is to show that widely accepted statements found in popular science books, textbooks, encyclopedias, and scholarly articles do not support detailed and rigorous analysis. To achieve this goal, this book does not offer an idealized view of electrodynamics, but rather a realistic view, including its main problems, inconsistencies, and limitations.

This book is intended not only for physicists and physics students, but also for engineers and other scientists, or simply curious people with a background in physics. This is not a book to learn electrodynamics for the first time, but a book to improve your knowledge of electrodynamics. This book is the first in a series devoted to common misconceptions in physics.

Juan R. González Álvarez
Vigo, February 2024.

—1

INTRODUCTION

Science is a dialogue with Nature, but although Nature is unique, science has traditionally been divided into large blocks. Scientists, philosophers, and librarians often talk about formal sciences, physical sciences, life sciences, and social sciences, with each of those blocks, in turn, is subdivided into smaller sciences.

The physical sciences focus on the study of the inorganic world, while living things are studied in the life and social sciences. The four main physical sciences are astronomy, physics, chemistry, and Earth sciences. For the last two, some authors speak instead of astronomical sciences and chemical sciences.

These divisions and subdivisions of science have the drawback that the formalism and methods used in nuclear physics are different from those used in anatomy, for example. Of course, there are interdisciplinary sciences such as biophysics and quantum chemistry that build bridges between disciplines, but these bridges are reduced to applying the formalism and methods of one discipline to the object of study of the other.

In fact, Zhou begins his question-and-answer session by defining biophysics[1] as "*the study of biological systems and biological processes using physics-based methods or based on physical principles*", while Levine states in the first chapter of his famous textbook[2] that "*Quantum chemistry applies quantum mechanics to problems in chemistry*". The important point here is that biophysics is not a true unification of biology and physics, and quantum chemistry does not unify quantum mechanics and chemistry.

Not only is there no true unification in these interdisciplinary sciences, but they are internally inconsistent in the sense that they combine elements from unrelated disciplines without verifying their mutual compatibility. Consider quantum chemistry: the incompatibility between chemistry and quantum mechanics was highlighted by the Nobel Prize laureate Prigogine:[3]

> "*Quantum mechanics, in its orthodox form, corresponds to a deterministic time-reversible description. This is not so for chemistry. Chemical reactions correspond to irreversible processes creating entropy. That is, of course, a very basic aspect of chemistry, which shows that it is not reducible to classical dynamics or quantum mechanics.*"

Unfortunately, we not only find inconsistencies in the boundaries between traditional disciplines, we also find inconsistencies in the traditional disciplines. Feynman, Leighton, and Sands write in section 28-1 of the second volume of *The Feynman Lectures on Physics*:[4]

> "*Now we want to discuss a serious trouble—the failure of the classical electromagnetic theory. You can appreciate that there is a failure of all classical physics because of the quantum-mechanical effects. Classical mechanics is a mathematically consistent theory; it just doesn't agree with experience. It is interesting, though, that the classical theory of electromagnetism is an unsatisfactory theory all by*

> *itself. There are difficulties associated with the ideas of Maxwell's theory which are not solved by and not directly associated with quantum mechanics. You may say, 'Perhaps there's no use worrying about these difficulties. Since the quantum mechanics is going to change the laws of electrodynamics, we should wait to see what difficulties there are after the modification.' However, when electromagnetism is joined to quantum mechanics, the difficulties remain. So it will not be a waste of our time now to look at what these difficulties are. Also, they are of great historical importance. Furthermore, you may get some feeling of accomplishment from being able to go far enough with the theory to see everything—including all of its troubles."*

This is a serious problem. A scientific theory can be more or less accurate, and its scope more or less general, but internal consistency is a necessary condition for a theory to be satisfactory, because an internally inconsistent theory can predict one thing and its opposite.

An internally inconsistent theory is only permissible as a preliminary stage of research that will eventually be superseded by a fully consistent theory. If we want to develop a consistent and unified picture of Nature, then we must identify the difficulties, inconsistencies, and incompatibilities in current disciplines, because *Nature is unique and internally consistent.*

Popular science books written by physicists, physics textbooks, encyclopedias, and academic physics literature contain erroneous or easily misinterpreted statements about physical phenomena and concepts. Some of those erroneous claims are exclusive to a single author or a small group, but other erroneous claims are repeated by thousands of scientists, engineers, philosophers, and journalists.

In this book we will focus on the second group of claims because their widespread popularization prevents a fundamental understanding of Nature, but first I will explain what I mean by myths and misconceptions.

Misconception: A mistaken belief or idea that is incorrect because it is based on a faulty interpretation or understanding.

Myth: A traditional and common misconception.

This book is intended to be pedagogical and readable by a wide audience, but all the myths and misconceptions are technical in nature, so this book contains equations and other technical details that are essential to understanding why these widely accepted claims are not entirely true.

Classical electrodynamics —or classical electromagnetism— is not only one of the classical branches of physics, but it is at the root of many engineering applications: generators that convert mechanical energy into electromagnetic energy, motors that perform the reverse conversion, lighting systems, displays, fans, bread toasters, magnetic card readers, loudspeakers, electromagnetic sensors, magnetic resonance imaging scanners, antennas and waveguides, electromagnetic suspensions, and many more. Classical electrodynamics is, in its own right, a fundamental part of the electrical engineering syllabus.

Beyond its numerous applications, classical electrodynamics has served as a theoretical foundation for the development of modern branches of physics such as special relativity and quantum field theory. As Feynman *et al.* point out in the aforementioned lectures, "*the classical theory of electromagnetism is an unsatisfactory theory all by itself*" and, not surprisingly, many of its deficiencies have been inherited by these modern branches.

Therefore, it seems a good idea to start the series dedicated to common misconceptions in physics with the analysis of the formalism and methods of classical electrodynamics.

We will begin this journey into the innermost secrets of electromagnetic phenomena with a chapter on the theoretical minimum necessary to understand the rest of the book. This is the content of chapter 2.

Chapter 3 is dedicated to the ancient myth that the field model of electromagnetic interactions has solved the mysteries of the action-at-a-distance model, and we will continue the journey by explaining in chapter 4 how the Lagrangians and Hamiltonians found in textbooks are divergent (formally infinite) and lack an unequivocal physical interpretation.

In chapter 5, we will refute the common claim that experiments have confirmed that electromagnetic interactions are delayed, proving in the process that the instantaneous potentials derived from the Maxwell equations are fully compatible with special-relativistic causality. We also provide examples of common applications of instantaneous potentials in classical theory and atomic physics.

Any student is taught that the Lorentz force is the force acting on a point charge in the presence of the electromagnetic field, but we will see in chapter 6 how the Lorentz force is neither the force associated with the Hamilton equations of motion nor the force we have to use in a Newton-type equation of motion for arbitrary regimes of velocity and field strengths.

Even when the Lorentz force is used in the proper equation of motion, it does not completely describe the motion of accelerating charged particles since this equation has to be supplemented with *ad hoc* radiation reaction forces. Chapter 7 is dedicated to delving into the concept of electromagnetic force and the main difficulties with radiation reaction.

Since classical electrodynamics has serious difficulties in describing the motion of a single charge, it is not surprising that this classical branch of physics cannot really explain the motion of many-body systems. Chapter 8 explains why the so-called relativistic field theory cannot describe many-body systems, except within certain limits.

Finally, the first appendix contains a detailed derivation of renormalized field theory from a theory of action at a distance, the second appendix shows why bosons can have speed but not velocity, and the last appendix contains a list of all the misconceptions discussed in this book.

—2

THE THEORETICAL MINIMUM

There are many excellent textbooks on electrodynamics. We have the second volume of *The Feynman Lectures On Physics*,[4] the introductory textbook by Griffiths,[5] the classic treatises by Jackson,[6] and by Landau and Lifshitz,[7] and those by Greiner[8] and Schwinger et al.[9] We can count three Nobel Prize laureates in physics on this list, with Feynman and Schwinger awarded for their contributions to quantum electrodynamics.

We will use Griffiths and Jackson for this chapter. Griffiths states in the opening pages of his textbook that electrodynamics is "*a beautifully complete and successful theory*", which has become something of a paradigm for physicists: "*an ideal model that other theories strive to emulate*". Beauty is a subjective term and I will not comment on that part, but we saw in the previous chapter how Feynman, Leighton, and Sands admit that the classical theory of electromagnetism is an unsatisfactory theory, and if we consult the textbook by Schwinger et al., we will see how the four authors affirm that classical electrodynamics is not yet a closed subject.

I think we can use Griffiths' words as an excellent example of what we mentioned in the preface: some physicists hide the main problems and shortcomings of physics from students and the general public. In fact, if we follow the main sources of science news and if we also read the most popular science books, we get the false impression that there are only a couple of unsolved mysteries in cosmology and quantum physics and that the rest of physics is complete and perfectly understood.

Griffiths writes:[5]

> *"The laws of classical electrodynamics were discovered in bits and pieces by Franklin, Coulomb, Ampere, Faraday, and others, but the person who completed the job, and packaged it all in the compact and consistent form it has today, was James Clerk Maxwell. The theory is now a little over a hundred years old."*

The Maxwell equations, sometimes called the Maxwell-Heaviside equations, are a set of four coupled partial differential equations

$$\nabla \cdot \boldsymbol{E} = \frac{\rho}{\epsilon_0}, \tag{2.1}$$

$$\nabla \cdot \boldsymbol{B} = 0, \tag{2.2}$$

$$\nabla \times \boldsymbol{E} + \frac{\partial \boldsymbol{B}}{\partial t} = \boldsymbol{0}, \tag{2.3}$$

and

$$\nabla \times \boldsymbol{B} - \frac{1}{c^2}\frac{\partial \boldsymbol{E}}{\partial t} = \mu_0 \boldsymbol{J}. \tag{2.4}$$

These equations are coupled because they cannot be solved independently, since knowing the value of certain quantities that appear in some equations depends on knowing the value of quantities that appear in the other equations. For example, we cannot solve equation (2.3) to obtain $\boldsymbol{E}$, without first knowing $\boldsymbol{B}$.

The quantities $\boldsymbol{E}$, $\boldsymbol{B}$, and $\boldsymbol{J}$ are three-dimensional vectors; all other quantities are scalars. $\boldsymbol{E}$ is the so-called electric field and $\boldsymbol{B}$ the so-called magnetic field, ρ is the electric charge density, and ϵ_0 and μ_0 are the permittivity and permeability of vacuum, respectively. $\boldsymbol{J}$ is the electric current density, this is the amount of charge per unit time that flows through a unit area (see also figure 2.1), and c is the speed of light in vacuum. The symbol ∇ denotes the three-dimensional gradient operator, the dot $\cdot$ is a scalar product between two vectors, and $\times$ is the cross product, also called the vector product. Finally, $\partial/\partial t$ is a partial derivative with respect to time.

The Maxwell equations take this form with the *International System of Units*. The equations are slightly different in other systems of units, but of course they describe the same physics.

Note: The vacuum, also called free space, is a concept used in classical field theory to refer to regions of space devoid of any particle. The term is a misnomer because it does not represents a true vacuum, since this theory assumes that those regions of space are filled with the electromagnetic field. Some authors use a different convention for $\boldsymbol{J}$ and refer to it as the electric current. The Gauss law is sometimes called the Coulomb law.

The terminology that I have used for $\boldsymbol{E}$ and $\boldsymbol{B}$ is standard but misleading, because it gives the impression that there are two different fields. Jackson writes in this regard:[6]

> *"The almost independent nature of electric and magnetic phenomena disappears when we consider time-dependent problems. Time-varying magnetic fields give rise to electric fields and vice-versa. We then must speak of electromagnetic fields, rather than electric or magnetic fields."*

But to be strict, there is a single electromagnetic field that extends throughout all space, and not a collection of electromagnetic fields.

Permittivity, permeability, and the speed of light are universal constants related by the identity $\epsilon_0\mu_0c^2 = 1$. This means that the term $\mu_0\boldsymbol{J}$ in equation (2.4) can be also written as $\boldsymbol{J}/\epsilon_0c^2$ and equation (2.1) could just as well be written as $\nabla \cdot \boldsymbol{E} = \mu_0c^2\rho$. The usual convention is to use ϵ_0 in (2.1) and μ_0 in (2.4). Do not ask me why. I just do not know.

All the physical quantities in the Maxwell equations, except the three universal constants, are functions of space and time; that is, $\boldsymbol{E} = \boldsymbol{E}(\boldsymbol{r}, t)$, $\boldsymbol{B} = \boldsymbol{B}(\boldsymbol{r}, t)$, $\rho = \rho(\boldsymbol{r}, t)$, and $\boldsymbol{J} = \boldsymbol{J}(\boldsymbol{r}, t)$, where $\boldsymbol{r}$ is the position vector.

Maxwell unified electric and magnetic phenomena within a single theory, but from a historical point of view, equation (2.1) is the Gauss law, equation (2.2) describes the absence of free magnetic poles, (2.3) is the Faraday law of induction, and (2.4) is the Ampère law with Maxwell's correction (the original law derived by Ampère was $\nabla \times \boldsymbol{B} = \mu_0\boldsymbol{J}$).

We can derive several interesting results from the Maxwell equations. For example, if we multiply (2.4) by the operator ∇, we use the mathematical identity $\nabla\cdot\nabla\times = 0$ and equation (2.1), we obtain the law of conservation of charge

$$\frac{\partial \rho}{\partial t} = -\nabla \cdot \boldsymbol{J}. \tag{2.5}$$

This law establishes that electric charges are neither created nor destroyed; that is, a charge cannot disappear at one point in space and appear instantaneously at another point, without flowing from the first point to the second.

Using this conservation law, we can show that the total electric charge $Q = \int \rho dV$ in a volume element remains constant unless charge enters or leaves the volume through its boundary, as shown in figure 2.1.

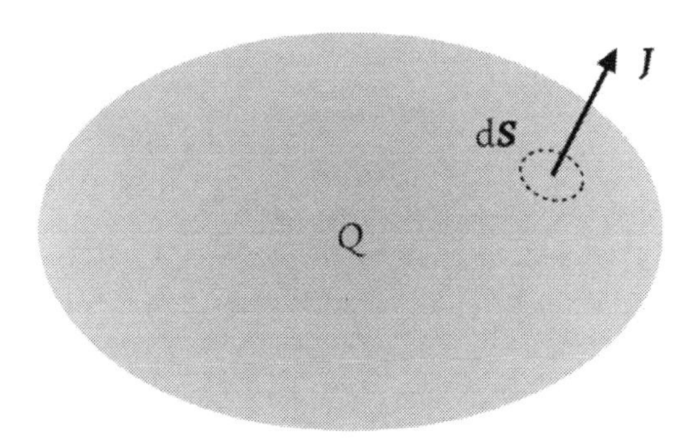

Figure 2.1: The change in the amount of total charge Q in a volume V is due to the net flow $\int \boldsymbol{J} \cdot \mathrm{d}\boldsymbol{S}$ of charge that enters or leaves the volume through its boundaries, with $\mathrm{d}\boldsymbol{S}$ the vector representing the infinitesimal area element.

With the Maxwell equations at hand, we could ask: given ρ and $\boldsymbol{J}$, what is the value of $\boldsymbol{E}$ and $\boldsymbol{B}$? In the static case, the Coulomb law and the Biot-Savart law provide the answer, but what is the solution for arbitrary time-dependent configurations of charge and current densities? Solving the Maxwell equations is not an easy task, but the problem is greatly simplified when we introduce the concept of electromagnetic potentials.

Without further ado, we introduce the vector potential $\boldsymbol{A}$ using the expression

$$\boldsymbol{B} = \nabla \times \boldsymbol{A}, \tag{2.6}$$

and introduce the scalar potential Φ, which is related to the electric field $\boldsymbol{E}$ by the following expression

$$\boldsymbol{E} = -\nabla\Phi - \frac{\partial \boldsymbol{A}}{\partial t}. \tag{2.7}$$

Using these relationships between fields and potentials, the four Maxwell equations for $\boldsymbol{E}$ and $\boldsymbol{B}$ can be reduced to the following pair of equations for the scalar and vector potentials

$$\nabla^2\Phi + \frac{\partial}{\partial t}(\nabla \cdot \mathbf{A}) = -\frac{\rho}{\epsilon_0} \tag{2.8}$$

and

$$\nabla^2\mathbf{A} - \frac{1}{c^2}\frac{\partial^2\mathbf{A}}{\partial t^2} - \nabla\left(\nabla \cdot \mathbf{A} + \frac{1}{c^2}\frac{\partial\Phi}{\partial t}\right) = -\mu_0\boldsymbol{J}. \tag{2.9}$$

This pair of coupled equations are called the Maxwell equations in the potential formulation and they contain all the information contained in the original Maxwell equations, but are easier to solve. Given ρ and $\boldsymbol{J}$, we can calculate Φ and $\boldsymbol{A}$ from (2.8) and (2.9), and then obtain $\boldsymbol{E}$ and $\boldsymbol{B}$. Evidently, the electromagnetic potentials are also functions of space and time: $\Phi = \Phi(\boldsymbol{r}, t)$ and $\boldsymbol{A} = \boldsymbol{A}(\boldsymbol{r}, t)$.

Different potentials can generate the same electromagnetic field. The potentials $\boldsymbol{A}$ and $\boldsymbol{A}' = \boldsymbol{A} + \nabla\lambda$, where λ is an arbitrary scalar function, produce the same $\boldsymbol{B} = \nabla \times \boldsymbol{A}' = \nabla \times (\boldsymbol{A} + \nabla\lambda) = \nabla \times \boldsymbol{A}$, because $\nabla \times \nabla\lambda = 0$ for any choice of λ. Likewise, the potentials Φ and $\Phi' = \Phi - (\partial\lambda/\partial t)$ produce the same $\boldsymbol{E}$.

This partial freedom in the choice of potentials implies that the above pair of equations for the scalar and vector potentials are not sufficient to fix the potentials. A third equation is needed: the choice of gauge. Different gauges are in use. Two common choices are the Lorenz and Coulomb gauges.

If we choose the Lorenz gauge, which is defined in SI units by

$$\nabla \cdot \mathbf{A}^{\text{Lor}} + \frac{1}{c^2}\frac{\partial\Phi^{\text{Lor}}}{\partial t} = 0, \tag{2.10}$$

then equations (2.8) and (2.9) are decoupled as

$$\nabla^2 \Phi^{\text{Lor}} - \frac{1}{c^2}\frac{\partial^2 \Phi^{\text{Lor}}}{\partial t^2} = -\frac{\rho}{\epsilon_0} \tag{2.11}$$

and

$$\nabla^2 \boldsymbol{A}^{\text{Lor}} - \frac{1}{c^2}\frac{\partial^2 \boldsymbol{A}^{\text{Lor}}}{\partial t^2} = -\mu_0 \boldsymbol{J}, \tag{2.12}$$

and can be solved independently. Furthermore, the two equations have a similar mathematical structure and we can use essentially the same analytical or numerical methods to solve them.

Note: The Lorenz gauge is named in honor of the Danish physicist Ludvig V. Lorenz, who is often confused with the Dutch physicist Hendrik A. Lorentz. This confusion is the reason why this gauge is called the "*Lorentz gauge*" in much of the literature.[6–9] Wald's recent book[10] is the only one here that uses the correct term. Griffiths mentions[5] in a footnote that "*here is some question whether this [gauge] should be attributed to H. A. Lorentz or to L. V. Lorenz*" and then justify the use of the wrong term because "*all the standard textbooks include the t, and to avoid possible confusion I shall adhere to that practice.*" The condition (2.10) was first published by L. V. Lorenz in 1867 and Maxwell cited his work. H. A. Lorentz rederived (2.10) in 1892, a year after the death of L. V. Lorenz. This rederivation is probably why many have mistakenly identified him with this gauge. I cannot understand Griffiths' position on this issue. Imagine what academia would be if we allowed errors and misconceptions to propagate forever in time!

If we chose the Coulomb gauge, defined by the condition

$$\nabla \cdot \boldsymbol{A}^{\text{Coul}} = 0, \tag{2.13}$$

then equations (2.8) and (2.9) reduce to

$$\nabla^2 \Phi^{\text{Coul}} = -\frac{\rho}{\epsilon_0} \tag{2.14}$$

and

$$\nabla^2 \boldsymbol{A}^{\text{Coul}} - \frac{1}{c^2}\frac{\partial^2 \boldsymbol{A}^{\text{Coul}}}{\partial t^2} - \frac{1}{c^2}\frac{\partial}{\partial t}(\nabla \Phi^{\text{Coul}}) = -\mu_0 \boldsymbol{J}. \tag{2.15}$$

The Maxwell equations (in both their original and potential form) allow us to calculate the electromagnetic field when charge and current densities are specified, and they also allow us to calculate the densities when the electromagnetic field is specified.

Of course, like any partial differential equation, the Maxwell equations must be supplemented with suitable boundary conditions, such as that $\boldsymbol{E}$ and $\boldsymbol{B}$ going to zero at large distances from a localized charge distribution. However, even considering boundaries, the Maxwell equations do not provide a complete theory of electromagnetism. The equations cannot be used to study the motion of charges in an electromagnetic field. For this last type of problem we need an equation of motion and the corresponding force law: for example, the Lorentz force

$$\boldsymbol{F}^{\mathrm{Lor}} = q(\boldsymbol{E} + \boldsymbol{v} \times \boldsymbol{B}), \tag{2.16}$$

where q is the electric charge and $\boldsymbol{v}$ its velocity with respect to a reference frame.

According to Griffiths,[5] the four Maxwell equations together the Lorentz force law "*summarize the entire theoretical content of classical electrodynamics*", except for some special properties of matter. Jackson makes a similar comment:[6] "*when combined with the Lorentz force equation and Newton's second law of motion, these equations provide a complete description of the classical dynamics of interacting charged particles and electromagnetic fields*".

Suppose we want to calculate the effect that a given electromagnetic field has on the movement of a charge, in what equation of motion would we use the Lorentz force law? Jackson mentions the Newton second law, but is he referring to the form $\boldsymbol{F} = m\boldsymbol{a}$ or the form $\boldsymbol{F} = (\mathrm{d}\boldsymbol{p}/\mathrm{d}t)$? The two equations are equivalent in Newtonian mechanics because the momentum $\boldsymbol{p}$ is related to the velocity by the simple relationship $\boldsymbol{p} = m\boldsymbol{v}$, with m the mass of the charge, but things are more complex in electrodynamics. Maybe we should use another equation of motion for the charge? To find the answer to these and other questions, we must resort to the science of mechanics.

Classical mechanics was born with Newton, but Hamiltonian and Lagrangian mechanics emerged as a reformulation of Newtonian mechanics and are now of great importance, since the new formulations of mechanics can be used to solve dynamical problem more easily and quickly than through Newtonian mechanics. The new formulations are based on the concepts of Hamiltonian and Lagrangian.

If we use Lagrangian dynamics to find the equation of motion, we need to know the Lagrangian L for a system of charges and the electromagnetic field. We can classify the main classical electrodynamics textbooks into those[5] that do not mention any Lagrangian, those[6] that give a Lagrangian valid only for low velocities (the so-called 'nonrelativistic' limit), and those texts[8] that give an incomplete Lagrangian where the electromagnetic field is missing. Landau and Lifshitz[7] prefer to give an action function in covariant notation, but we can derive a standard noncovariant Lagrangian from the action. The result is

$$L = -\sum_i^N m_i c^2 \sqrt{1 - \frac{\boldsymbol{v}_i^2}{c^2}} + \int (\boldsymbol{J} \cdot \boldsymbol{A} - \rho\Phi)\, \mathrm{d}V + \epsilon_0 \int \frac{\boldsymbol{E}^2 - c^2\boldsymbol{B}^2}{2}\, \mathrm{d}V. \tag{2.17}$$

The sum is over the total number N of charges, and the integrations are calculated over the total volume V that confines the charges and the electromagnetic field. Here m_i is the mass of each charged particle, $\boldsymbol{v}_i$ its velocity, and the rest of quantities have been defined before.

Note: Remember that we are using SI units in this book instead of Gaussian units. We are also using a definition for the vector potential that might differ from the definitions used by other authors. This means you may find slight differences in notation between formulae in this book and those elsewhere. For example, some authors write $(\boldsymbol{J}\cdot\boldsymbol{A}/c - \rho\Phi)$ for the middle term in (2.17), because they use a different definition for $\boldsymbol{A}$. In books that use Gaussian units, you will find $(1/8\pi)\int(\boldsymbol{E}^2 - \boldsymbol{B}^2)\,\mathrm{d}V$ in the last term of the aforementioned equation, with $\pi = 3.14159265358....$ Additionally, some authors use $\int(d\boldsymbol{r})$ or $\int d^3x$ to denote a volume integral. Consequently, the rest of formulae in this book should differ from those in other books, specially older books that use Gaussian units. Of course, the physics described is the same.

If finding the expression (2.17) in electrodynamics textbooks is a difficult task, finding the corresponding Hamiltonian H is almost impossible. Luckily, we can apply a Legendre transformation to (2.17) and obtain the following Hamiltonian for a system of charges and electromagnetic field

$$H = \sum_i^N \sqrt{m_i^2 c^4 + (\boldsymbol{p}_i - q_i\mathbf{A})^2 c^2} + \epsilon_0 \int \frac{\mathbf{E}^2 + c^2\mathbf{B}^2}{2}\,\mathrm{d}V, \tag{2.18}$$

where q_i is the electric charge of the particle and $\boldsymbol{p}_i$ its momentum. The rest of the symbols have the same meaning as described before. Using either the Hamiltonian or the Lagrangian we can obtain the equations of motion.

For future reference, we will approximate the above Hamiltonian for the physical situation when the charged particles are moving at low speeds; that is, when $|\boldsymbol{v}_i| \ll c$

$$H = \sum_i^N \frac{(\boldsymbol{p}_i - q_i\mathbf{A})^2}{2m_i} + \epsilon_0 \int \frac{\mathbf{E}^2 + c^2\mathbf{B}^2}{2}\,\mathrm{d}V. \tag{2.19}$$

The rest energy term $\sum_i^N m_i c^2$ has been dropped in the above Hamiltonian, as it is a mere constant and does not affect the equations of motion.

We have to our disposal the basic equations and concepts involved in the field theory of electromagnetism, but we need a framework of interpretation to make sense of it all. What is this framework? In essence, the Maxwell equations are needed to obtain ***E*** and ***B*** from the distribution of charge densities and currents, or vice versa, and ***E*** and ***B*** are then introduced into the Lorentz force law that must be used in a specific equation of motion for the charges, as will see in a posterior chapter.

This basic framework seems pretty clear, even if we have not studied electrodynamics before and have only taken a look at the Maxwell-Lorentz equations. However, we may have other fundamental questions; for example, why fields and charges? Why not just charges like in the original Coulomb theory? Griffiths writes that the fundamental problem that a theory of electromagnetism hopes to solve is the following:[5]

> *"I hold up a bunch of electric charges here (and maybe shake them around)–what happens to some other charge, over there? The classical solution takes the form of a field theory: We say that the space around an electric charge is permeated by electric and magnetic fields (the electromagnetic 'odor,' as it were, of the charge). A second charge, in the presence of these fields, experiences a force; the fields, then, transmit the influence from one charge to the other-they mediate the interaction."*

Jackson offers an extended discussion:[6]

> *"Although* **E** *and* **B** *thus first appear just as convenient replacements for forces produced by distributions of charge and current, they have other important aspects. First, their introduction decouples conceptually the sources from the test bodies experiencing electromagnetic forces. If the fields* **E** *and* **B** *from two source distributions are the same at a given point in space, the force acting on a test charge or current at that point will be the same, regardless of*

> *how different the source distributions are. This gives* **E** *and* **B** *in (I.3) meaning in their own right, independent of the sources. Second, electromagnetic fields can exist in regions of space where there are no sources."*

So far we have omitted an important consideration: what is a field? Griffiths tries to answer this question:

> *"But I encourage you to think of the field as a 'real' physical entity, filling the space in the neighborhood of any electric charge. Maxwell himself came to believe that electric and magnetic fields represented actual stresses and strains in an invisible primordial jellylike 'ether'. Special relativity has forced us to abandon the notion of ether, and with it Maxwell's mechanical interpretation of electromagnetic fields. [...] I can't tell you, then, what a field is—only how to calculate it and what it can do for you once you've got it."*

So, as ironic as it may seem, electromagnetic field theory is based on fields, but field theorists cannot tell us what they really are.

With the framework of interpretation and the concepts and equations that we have introduced, we are ready to begin reviewing the main misconceptions found in the literature.

—3

ACTION AT A DISTANCE

The action-at-a-distance model began in 1684 with Newton's theory of gravitation and the Newtonian expression for the force of attraction between two point masses separated by a nonzero distance. This model was very satisfactory since it accurately described the motion of astronomical objects. Although Newtonian gravity cannot describe phenomena associated with very massive bodies and high speeds, astronomers continue to use Newtonian gravity for those situations where it continues to provide accurate predictions.

But before continuing with the exposition, it is very important to point out that the term "*action-at-a-distance*" is a bit of a misnomer, because it gives us the impression that it is a model of interactions where a free body produces an influence that travels across space until it reaches another free body and carries this action from one body to the other. However, there are no intermediaries in this model and the whole idea of a traveling influence becomes absurd.

Newton, in a well-known letter that he wrote in February of 1693 to Bentley, was deeply cautious of the idea of a traveling influence:

> *"That one body may act upon another at a distance through a vacuum, without the mediation of anything else, by and through which their action and force may be conveyed from one to the other, is to me so great an absurdity, that I believe no man who has in philosophical matters a competent faculty of thinking, can ever fall into it."*

Field theorists like the Nobel Prize laureate Wilczek claim that "*Newton was extremely unhappy with one of his most glorious discoveries*"[11] and take the aforementioned letter as evidence that Newton rejected the notion of action at a distance. However, it has been argued that Newton does not say that this action is inconceivable and much less that it is impossible. Historians claim that Newton only rejected the intervention of a *material* substance between the gravitating masses.

To understand Newton's views on action at a distance, we need to study not only his letter to Bentley, but also all his manuscripts and books, including the "*Queries*" that were published in a later edition of *The Opticks*, in which Newton frequently discussed action at a distance (the first edition of his book, published in 1704, contained only sixteen queries; therefore, we will refer to the fourth edition of 1730).

In "*Query 28*", Newton rejects a mechanical medium to explain the motion of the planets and comets:

> *"And for rejecting such a Medium we have the authority of those the oldest & most celebrated Philosophers of Greece & Phoenicia, who made a Vacuum, Atoms & the gravity of Atoms the first principles of their philosophy."*

And in "*Query 31*", he writes:

> "*Have not the small Particles of Bodies certain Powers, Virtues or Forces, by which they act at a distance, not only upon the Rays of Light for reflecting, refracting and reflecting them, but also upon one another for producing a great part of the Phænomena of Nature? For it's well know that Bodies act one upon another by the Attractions of Gravity, Magnetism and Electricity.*"

Newton conceived two mechanisms for the action of forces: either movement occurs without the intervention of a substance or it occurs through the intervention of an *immaterial* substance.[12] Regardless of what is the correct interpretation of Newton's works, our understanding of the physical world has come a long way since he passed away. So how about we ignore what was written more than three centuries ago and focus on what we know now?

The old Newtonian picture in which each body or particle is considered a separate unit that interacts with other units by means of forces differs from the modern perspective of analytical mechanics. The distinction between the views of the vectorial mechanics developed by Newton and the analytical mechanics developed by Lagrange and Hamilton was vigorously emphasized by Lanczos in the first chapter of his treatise:[13]

> "*The analytical form of mechanics, as introduced by Euler and Lagrange, differs considerably in its method and viewpoint from vectorial mechanics. The fundamental law of mechanics as stated by Newton: 'mass times acceleration equals moving force' holds in the first instance for a single particle only.*
>
> *If the particle is not free but associated with other particles, as for example in a solid body, or a fluid, the Newtonian equation is still applicable if the proper precautions are observed. One has to isolate the particle from all other particles and determine the force which is exerted on it by the surrounding particles. Each particle*

is an independent unit which follows the law of motion of a free particle.

The analytical approach to the problem of motion is quite different. The particle is no longer an isolated unit but part of a 'system'. A 'mechanical system' signifies an assembly of particles which interact with each other. The single particle has no significance; it is the system as a whole which counts. For example, in the planetary problem one may be interested in the motion of one particular planet. Yet the problem is unsolvable in this restricted form. The force acting on that planet has its source principally in the sun, but to a smaller extent also in the other planets, and cannot be given without knowing the motion of the other members of the system as well. And thus it is reasonable to consider the dynamical problem of the entire system, without breaking it into parts."

When a mechanical system contains many particles, the vector approach is very difficult to apply. The analytical approach simplifies mechanical problems by treating such systems as a collection of particles plus the network of their mutual interactions, rather than considering each particle as a unit subject to external forces (see figure 3.1 for a comparison of both approaches).

Note: Because of this difference between ancient and modern pictures, and to avoid further confusions regarding the concept of traveling influences, some scientists prefer to use the terms *direct-action* and *direct-interparticle-action* instead of *action-at-a-distance*. The new terminology is not without problems, and here and the following we will retain, for reasons of historical consistency, the old term.

Analytical mechanics uses certain scalar properties that represent the system as a whole, instead of vector forces for each particle, to analyze the motion of the system. One of those properties is the potential energy U, a function that contains all the necessary information about the interactions, including their strength and range.

For a system with N particles, the total energy E is equal to the sum of the energy E_i of each particle plus the potential energy of the system: $E = \sum_i^N E_i + U$.

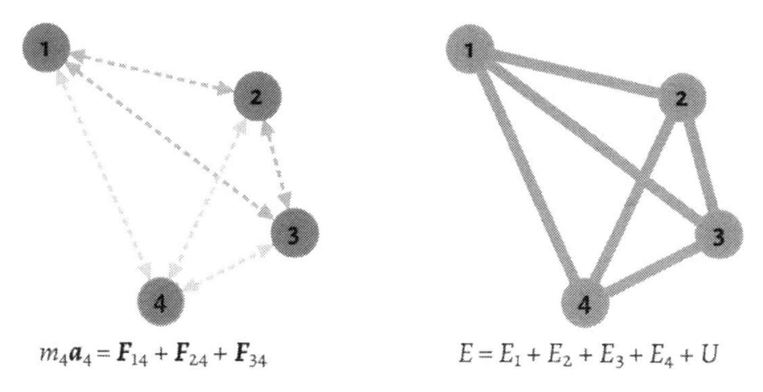

$m_4\boldsymbol{a}_4 = \boldsymbol{F}_{14} + \boldsymbol{F}_{24} + \boldsymbol{F}_{34}$ $\qquad$ $E = E_1 + E_2 + E_3 + E_4 + U$

Figure 3.1: Vectorial and analytical pictures of the same mechanical system. In the vectorial picture (left) we have a collection of individual particles plus the external forces acting on each pair of particles, and we apply the equation of motion to each particle (we illustrate the case of particle number 4). In the analytical picture (right) we have a single system defined by the collection of particles plus the network of interactions that define the system as a whole.

In the action-at-a-distance model, causality arises not as an influence that propagates with a finite speed between isolated particles, but through the evolution of the state of the system as a whole.

About a century after Newton's discoveries, Coulomb was able to measure the force between electrified bodies, using the torsion balance he invented. Imitating Newton, Coulomb explained the experiments in terms of an action-at-a-distance formula for the interaction between charges; this formula is now called the Coulomb force. It was not until the beginning of the 19th century that Ampère described all then known electromagnetic phenomena in a unified way, in terms of an action-at-a-distance model.

About a decade later, Faraday, unable to explain the recent experimental discoveries (especially his law of induction) in terms of the Coulomb-Ampère model, considered in 1849 reinterpreting all electromagnetic phenomena in terms of an alternative model based on something new he called "*fields*". According to Faraday, the magnetic field was a region of the space close to magnetized bodies and the electric field was a region of the space close to electrified bodies.

Maxwell's subsequent work laid the foundation for the field approach to classical electrodynamics in 1865, while Hertz's experimental observation of electromagnetic waves in 1887 *appeared* to confirm the field concept.

Since the nineteenth century, we can find in the literature the false claims that "*classical electrodynamics is a field theory*"[14] or that "*classical field theory and experiment imply fields are fundamental, and indeed Faraday, Maxwell, and Einstein concluded as much*".[15] Nothing could be further from the truth! Not only does the classical field model have a dual nature in which the charges are described as particles and not fields, but Wheeler and Feynman demonstrated in their paper *Classical Electrodynamics in Terms of Direct Interparticle Action*[16] that all electromagnetic phenomena described by the Faraday-Maxwell theory can be understood without using fields.

Note: The Faraday-Maxwell formulation in term of fields is traditionally called the *contact-action* model of the interactions, but this is another misnomer, because it gives the false belief that action-at-a-distance models cannot describe interactions by contact when, in fact, the same formula that explains the interactions between two electrons on opposite sides of our galaxy can explain the interaction between the electrons in our fingers and the electrons on the surface of a wooden table. We will refer to the Faraday-Maxwell formulation of interactions as the field model of interactions. As the previous paragraph might have suggested to us, the popular term *classical field theory* is another misnomer, because this theory cannot explain electromagnetic phenomena without considering charged particles in addition to the electromagnetic field.

Critics of the action-at-a-distance model of electromagnetic interactions have traditionally argued that this model does not explain how charges can interact directly without mediators. If we use the symbol **?** to denote the existence of a mystery, then this model can be represented formally as the following direct coupling

$$\boldsymbol{charge} \overset{?}{\Longleftrightarrow} \boldsymbol{charge}. \tag{3.1}$$

The aforementioned critics propose replacing the action-at-a-distance model by another where a field mediates the electromagnetic interactions between charges

$$\boldsymbol{charge} \overset{?}{\Longleftrightarrow} \boldsymbol{field} \overset{?}{\Longleftrightarrow} \boldsymbol{charge}, \tag{3.2}$$

but those critics have replaced the original mystery in (3.1) with new mysteries: How do charges and field interact without any mediator between the field and the charges? We are in a situation worse than the one represented in (3.1), and if we were to introduce additional intermediaries **X** between the charges and the field, we would only increase the number of mysteries

$$\boldsymbol{charge} \overset{?}{\Longleftrightarrow} \boldsymbol{X} \overset{?}{\Longleftrightarrow} \boldsymbol{field} \overset{?}{\Longleftrightarrow} \boldsymbol{X} \overset{?}{\Longleftrightarrow} \boldsymbol{charge}. \tag{3.3}$$

Therefore, classical field theorists simply postulate a direct coupling (3.2) between the field and the charges, and do not provide any further explanation of what the supposed physical mechanism of that coupling is. Replacing the action-at-a-distance model (3.1) with the field model (3.2) has not increased our understanding of the electromagnetic interaction.

Note: Quantum field theorists use a more elaborate model, in which one charge emits a virtual photon (sometimes called a *force carrier*) that crosses space until it is absorbed by the other charge. However, quantum field theory does not provide any physical mechanism for the supposed emission and absorption of virtual particles, nor does it explain how the virtual photon knows in advance which direction it must travel in order to later encounter the other charge.

The fact that classical field theory has not solved any mysteries is also revealed when the details of the interactions in each of the models are analyzed.

In the action-at-a-distance model, the Coulomb potential energy associated with two charges q_1 and q_2 placed at positions $\boldsymbol{r}_1$ and $\boldsymbol{r}_2$ is given by the expression

$$U^{\text{Coul}} = \kappa \frac{q_1 q_2}{|\boldsymbol{r}_1(t) - \boldsymbol{r}_2(t)|}, \tag{3.4}$$

with $\kappa = 1/(4\pi\epsilon_0)$ the Coulomb constant, also called the electric force constant or electrostatic constant, which measures the strength of the interaction. The vector $\boldsymbol{r}_1(t) - \boldsymbol{r}_2(t)$ is the distance between the charges and the vertical bars in (3.4) denote the magnitude of this vector.

This expression for the energy can be simplified if we define the Coulomb potential for the charge i

$$\phi_i = \kappa \frac{q_i}{|\boldsymbol{r}_1(t) - \boldsymbol{r}_2(t)|}. \tag{3.5}$$

Using this definition, the potential energy for the system of two charges can be written as

$$U^{\text{Coul}} = q_1 \phi_2 \tag{3.6}$$

or as

$$U^{\text{Coul}} = q_2 \phi_1. \tag{3.7}$$

The equality $q_1\phi_2 = q_2\phi_1$ is a consequence of the symmetries of the interaction in the action-at-a-distance model, because the energy (3.4) describes a true two-body interaction where both charges are on equal footing (we will see that this is not true in the field model). Note that there is a Coulomb potential for each charge and that these potentials are just a notational shorthand to simplify some formulas.

What is the alternative to (3.6) and (3.7) in classical field theory? Before giving an answer, I want to mention that we will see in chapter 4 that, after more than a century, physicists still do not agree on the physical interpretation of the potential energy or what its more general expression is. We will also see in chapter 8 that the field model cannot completely describe the interactions in a system of charges.

For the purposes of the present chapter, it will be sufficient to use the following standard expression[6] for the potential energy of a particle i coupled to the electromagnetic field

$$U^{\text{Maxw}} = q_i \Phi, \tag{3.8}$$

where Φ is the scalar potential that was introduced by equation (2.7). The symmetries of the original Coulomb interactions are lost in the field model because $q_1\Phi \neq q_2\Phi$, which has serious consequences for the development of a many-body theory, as we will see in chapter 8.

The action-at-a-distance model treats all particles symmetrically, and this is the reason why the energy (3.4) does not change if we swap particles 1 and 2. However, the field model makes an artificial distinction between the *source particles* that generate the field and the *test particles* that detect the field (also called "*source charges*" and "*test charges*" in the context of electromagnetic theory).[5] In the expression $q_1\Phi$, the particle q_1 is the test charge while that q_2 is a the source. The roles for q_1 and q_2 are reversed in $q_2\Phi$.

Note: I am using different symbols for the action-at-a-distance potential and for the field potential, but the same symbol is often used in the literature for both, which causes a lot of confusion and problems. We will see later how most of the literature calls Φ the Coulomb potential, although it is neither ϕ_1 nor ϕ_2.

Working in the Coulomb gauge for simplicity, the scalar potential given by field theory is

$$\Phi = \kappa \left(\frac{q_1}{|\boldsymbol{r} - \boldsymbol{r}_1(t)|} + \frac{q_2}{|\boldsymbol{r} - \boldsymbol{r}_2(t)|} \right). \tag{3.9}$$

We will ignore for now that for a two-particle system the Coulomb potentials are nonlocal and time-implicit functions $\phi_i(\boldsymbol{r}_1(t), \boldsymbol{r}_2(t))$ defined in an abstract configuration space of six dimensions (three coordinates per particle), while the potential of the classical field theory is a local and time-explicit function $\Phi(\boldsymbol{r}, t)$ defined in a four-dimensional spacetime. We will also ignore for now that the potentials of the action-at-a-distance theory are finite and with direct physical meaning while the potential of the classical field theory is a divergent quantity that must be corrected by a renormalization procedure. What we want to emphasize now is that the classical-field expression (3.8) has not solved the mystery of how charges interact.

Effectively, if we make the substitutions

$$q_1\phi_2 \longrightarrow q_1\Phi \tag{3.10}$$

and

$$q_2\phi_1 \longrightarrow q_2\Phi \tag{3.11}$$

in the equations of motion or in the expression for the energy, we do not obtain a more detailed understanding of the nature of the interaction. Both before and after the replacement we have the product of a charge and a potential. In the action-at-a-distance model, this product represents the coupling between charges, while in the field model the product represents the coupling between field and charge. If we evaluate the physical and mathematical properties of each potential, we do not find any new physics in Φ that was not present in ϕ_1 or in ϕ_2.

In fact, we will see in appendix A how the field model and its potentials Φ and $\boldsymbol{A}$ can be obtained from the action-at-a-distance model as a local approximation.

I mentioned earlier that the scalar potential of the field theory is a divergent quantity, we can strictly write that $\Phi = \infty$. The potential (3.9) diverges when the position $\boldsymbol{r}$ where the field intensity is evaluated coincides with the position of one of the charges. The reason for the presence of divergences can be easily understood by comparing the diagrams for each model of the electromagnetic interaction. The diagram (3.1) for the action-at-a-distance model can be decoupled into the partial diagram

$$\textbf{\textit{charge}} \xRightarrow{?} \textbf{\textit{charge}}, \tag{3.12}$$

representing the interaction exerted by the first particle on the second, and the partial diagram

$$\textbf{\textit{charge}} \xLeftarrow{?} \textbf{\textit{charge}}, \tag{3.13}$$

representing the interaction exerted by the second particle on the first.

The introduction of the electromagnetic field as an intermediary for the interaction between charges has negative consequences, because the diagram (3.2) is now decoupled into the following partial diagrams

$$\overset{\curvearrowleft\,?}{\textbf{\textit{charge}} \xRightarrow{?} \textbf{\textit{field}}} \xRightarrow{?} \textbf{\textit{charge}} \tag{3.14}$$

and

$$\textbf{\textit{charge}} \xLeftarrow{?} \underset{\curvearrowright\,?}{\textbf{\textit{field}} \xLeftarrow{?} \textbf{\textit{charge}}}. \tag{3.15}$$

In the top diagram, diagram (3.14), both charges 'feel' the electromagnetic field, but only the charge on the left side is the source of the field. The bent arrow means that the charge interacts with its own field; that is, the particle on the left is both a source and a test charge. This self-field is divergent because the distance between the source charge and the test charge is zero, since they are both the same particle. The roles of the particles are swapped in the bottom diagram, diagram (3.15), with the charge on the right side now interacting with its own field. These divergences are the reason why the potential energy (3.8) is not physical, but $U^{\text{Maxw}} = \infty$.

The interaction of a charged particle with itself is a nonphysical process unique to the field theory of electromagnetism. By introducing a field as an intermediary, the 'coupling' of the charge with the electromagnetic field can be transmitted from the field to another charge or back to the original charge.

We could introduce a much more detailed and technical analysis of both models of interactions, and we could even introduce elements of quantum field theory and discuss Feynman diagrams, propagators, and virtual particles, but the main conclusions would not change. The conclusion we can draw from this classical analysis of the electromagnetic interaction is that replacing the action-at-a-distance model of Newton and Coulomb with the field model of Faraday and Maxwell does not explain how particles interact, but rather introduces a more complex, nonphysical, and limited model of electromagnetic phenomena.

So why is the field model so popular? In the introduction to his textbook, Jackson defends field theory in the following terms:[6]

> *"In fact, though there are recurring attempts to eliminate explicit reference to the fields in favor of action-at-a-distance descriptions of*

> *the interaction of charged particles, the concept of the electromagnetic field is one of the most fruitful ideas of physics, both classically and quantum mechanically.*"

What Jackson does not tell us is that the classical theory of electromagnetic fields and charges has not been fully accepted by scientists and engineers because it is plagued by fundamental deficiencies: *violations of causality and of the law of inertia, divergences, the 4/3 problem, runaway pathologies, the Cullwick paradox, etc.* This list of difficulties and paradoxes is the reason why Feynman et al. admitted that "*the classical theory of electromagnetism is an unsatisfactory theory all by itself*". Furthermore, the list of deficiencies increases when a field is quantized, but this is material for another book (one volume in this series is dedicated to quantum field theory).

Griffiths states that "*it is even possible, though cumbersome, to formulate classical electrodynamics as an 'action-at-a-distance' theory, and dispense with the field concept altogether*",[5] but he does not explain why he finds this cumbersome.

There is renewed interest in action-at-a-distance models of interactions because these models provide a useful framework for constructing mathematical descriptions beyond the theoretical and computational limitations of classical and quantum field theories: from the dynamics of hadrons and nuclei at intermediate energies[17] to fundamental cosmological considerations.[18]

The main **conclusions** of this chapter are the following:

- Classical electrodynamics is not a field theory, because we can formulate it as an action-at-a-distance theory. Even the so-called field theory of electromagnetism is actually a theory of both charges and the electromagnetic field.

- Experiments do not imply that fields are fundamental elements of nature, since the same experiments can be explained without fields.
- In field theory there is a single electromagnetic field, not separate electric and magnetic fields.
- The field model does not solve the mystery of how charged particles interact. Furthermore, this model has introduced a long list of difficulties and paradoxes in physics.
- All electromagnetic phenomena described by the Faraday-Maxwell theory can be understood without using the concept of electromagnetic field.
- The terms "*action-at-a-distance*", "*contact-action*", and "*classical field theory*" are misnomers.

—4

LAGRANGIANS AND HAMILTONIANS

The central elements of Lagrangian and Hamiltonian mechanics are the Lagrangian and Hamiltonian functions, respectively. Both functions can be used to obtain the equations of motion of a system. In the context of field theory, the term "*equation of motion*" refers not only to the traditional equations of motion of the charged particles, but also to the Maxwell equations (2.1), (2.2), (2.3), and (2.4), which are considered to be the equations of motion of the electromagnetic field, since the field is taken to be a physical system in its own right.

We now begin our analysis of the standard Lagrangian (2.17) for a system of charges and electromagnetic field. We saw in the previous chapter that the scalar potential Φ is a divergent quantity due to the self-interaction of the charges with the field that they generate. The vector potential $\boldsymbol{A}$ also diverges for the same reason. This means that the Lagrangian is divergent, $L = \infty$, but physicists behave as if (2.17) were finite and physical.

If we want to gain a deeper understanding of the field theory of electromagnetism, we would ask ourselves a fundamental question: what is the physical interpretation of each term in the Lagrangian (2.17)?

It is very common[7] to consider the first term as the Lagrangian for "*free particles*" (the material system of charges), the last term as the Lagrangian for "*a field in the absence of charges*" (the free electromagnetic field), and the middle term as the "*interaction between the particles and the field*". In explicit form

$$L = \underbrace{-\sum_{i}^{N} m_i c^2 \sqrt{1 - \frac{\boldsymbol{v}_i^2}{c^2}}}_{\text{FREE CHARGES}} + \underbrace{\int (\boldsymbol{J} \cdot \boldsymbol{A} - \rho \Phi)\, \mathrm{d}V}_{\text{INTERACTION}} + \underbrace{\epsilon_0 \int \frac{\boldsymbol{E}^2 - c^2 \boldsymbol{B}^2}{2}\, \mathrm{d}V}_{\text{FREE FIELD}}. \tag{4.1}$$

Sometimes, the first two terms are identified with the system of charges[10]

$$L = \underbrace{-\sum_{i}^{N} m_i c^2 \sqrt{1 - \frac{\boldsymbol{v}_i^2}{c^2}} + \int (\boldsymbol{J} \cdot \boldsymbol{A} - \rho \Phi)\, \mathrm{d}V}_{\text{CHARGES}} + \underbrace{\epsilon_0 \int \frac{\boldsymbol{E}^2 - c^2 \boldsymbol{B}^2}{2}\, \mathrm{d}V}_{\text{FREE FIELD}}. \tag{4.2}$$

However, some physicists[6] combine the two integral terms into the "*Lagrangian for the electromagnetic field*", often also called[19] the "*Maxwell Lagrangian*", and consider the total Lagrangian (2.17) to be the sum of the terms of the charges and the electromagnetic field, as shown below,

$$L = \underbrace{-\sum_{i}^{N} m_i c^2 \sqrt{1 - \frac{\boldsymbol{v}_i^2}{c^2}}}_{\text{CHARGES}} + \underbrace{\int (\boldsymbol{J} \cdot \boldsymbol{A} - \rho \Phi)\, \mathrm{d}V + \epsilon_0 \int \frac{\boldsymbol{E}^2 - c^2 \boldsymbol{B}^2}{2}\, \mathrm{d}V}_{\text{FIELD}}. \tag{4.3}$$

The interpretation (4.1) is motivated by the following reasoning: if we remove the electromagnetic field by setting the potentials to zero for the entire space, then $\boldsymbol{E} = \boldsymbol{B} = \boldsymbol{0}$ and the remaining term in (4.1), the summation term with the square roots, corresponds to a system of free charges. The same line of reasoning suggests that we can remove the system of charges by setting the masses and charges to zero, which implies $\rho = 0$ and $\boldsymbol{J} = \boldsymbol{0}$, such that the remaining term in (4.1), the rightmost term, corresponds to a free electromagnetic field. But not everyone agree with this justification and this is why (4.2) and (4.3) exist.

The justification offered for the combination of the first two terms in (4.2) in the Lagrangian for charges is that both terms are needed to obtain *"the Lorentz force equation of motion of a charged particle in an electromagnetic field"*.[10]

As mentioned in the introduction to this chapter, the Maxwell equations are considered to be the equations of motion of the electromagnetic field. The alternative interpretation (4.3) is based precisely on the fact that to obtain the complete form of the Maxwell equations from the Lagrangian (2.17), we must use both integral terms and not only the last one.

What is the correct physical interpretation of the Lagrangian? None of the three interpretations stands up to close scrutiny. Consider (4.1) or (4.2), using the relationships between fields and potentials, we can show that the expression $\epsilon_0(\boldsymbol{E}^2 - c^2\boldsymbol{B}^2)$ contains a part proportional to $(\boldsymbol{J}\cdot\boldsymbol{A} - \rho\Phi)$ and cannot represent a free field in presence of charges. Furthermore, the last two terms in (4.3) also cannot represent the field, because those terms depend on matter variables, and a proper field term can only depend on field variables.

The Lagrangian (2.17) is valid for any gauge, but we can gain additional insight by working on the Coulomb gauge because then the scalar potential Φ no longer depends on the degrees of freedom of the field, which implies that the term $\rho\Phi$ cannot be associated with the electromagnetic field or with the interaction between the field and matter. The difficulties in interpreting this term are much better appreciated in the Hamiltonian formalism, as we will see in a moment.

As with the Lagrangian, the Hamiltonian of classical electrodynamics is also divergent, $H = \infty$, but physicists behave as if this Hamiltonian represented a physical energy. Since the Lagrangian and the Hamiltonian are related by a Legendre transformation, it should not surprise us that there is much disagreement about how to physically interpret each term in the Hamiltonian of a system of charges and the electromagnetic field.

Some physicists consider that the square root terms in the Hamiltonian (2.18) represent the combination of the kinetic and rest energies of matter (that is, giving the total energy of a system of free charges) and that the integral term represents the total energy of the electromagnetic field or the "*energy stored in electromagnetic fields*".[5] That is, they propose the following interpretation

$$H = \underbrace{\sum_i^N \sqrt{m_i^2 c^4 + (\boldsymbol{p}_i - q_i \mathbf{A})^2 c^2}}_{\text{FREE CHARGES}} + \underbrace{\epsilon_0 \int \frac{\mathbf{E}^2 + c^2 \mathbf{B}^2}{2} \mathrm{d}V}_{\text{FIELD}}. \tag{4.4}$$

But other physicists[9,10] do not agree with the previous interpretation, because an energy similar to that of Coulomb is missing in the system of charges. Therefore, they add the term $\sum_i^N q_i \Phi$ to the energy of the charges and then subtract this term, after transforming it into a volume integral, from the field term

$$H = \underbrace{\sum_i^N \sqrt{m_i^2 c^4 + (\boldsymbol{p}_i - q_i \mathbf{A})^2 c^2} + q_i \Phi}_{\text{CHARGES}} + \underbrace{\epsilon_0 \int \left(\frac{\mathbf{E}^2 + c^2 \mathbf{B}^2}{2} - \frac{\rho\Phi}{\epsilon_0} \right) \mathrm{d}V}_{\text{FREE FIELD}}. \tag{4.5}$$

Wald[10] uses (2.1) to replace the charge density with the gradient of the electric field because "*it is very important that the Hamiltonian be expressed as the correct function of its field variables to get the correct equations of motion*" (personal communication). The resulting interpretation is

$$H = \underbrace{\sum_i^N \sqrt{m_i^2 c^4 + (\mathbf{p}_i - q_i \mathbf{A})^2 c^2} + q_i \Phi}_{\text{CHARGES}} + \underbrace{\epsilon_0 \int \left(\frac{\mathbf{E}^2 + c^2 \mathbf{B}^2}{2} - \Phi \nabla \cdot \mathbf{E} \right) \mathrm{d}V.}_{\text{FREE FIELD}} \tag{4.6}$$

Schwinger et al.[9] work with a small velocity approximation in which the square roots are replaced with $\sum_i^N (\boldsymbol{p}_i - q_i \mathbf{A})^2 / 2m_i$ plus rest energy terms that are eliminated by energy rescaling. Of course, this approximation does not change the discussion about how they interpret each term in the Hamiltonian for a system of charges and electromagnetic field; they use (4.5).

The aforementioned authors state in section 9.3 of their book[9] that the ambiguities in the interpretation of the Hamiltonian are a consequence of "*whether the potential energy of particles is attributed to them or to the fields, or to both*". However, the situation is much more complicated than that. We can make at least three comments in this regard.

First, they identify the "*potential energy*" of the system of charges with $\sum_i^N q_i \Phi$, but the terms in the Hamiltonian (4.6) that depend on the vector potential $\boldsymbol{A}$ would also be considered part of the interaction, since those terms depend on the distance between the charges. Including the terms that depend on $\boldsymbol{A}$ in a generalized concept of potential energy is just what we must do to derive the Darwin potential energy in the classical theory[20,21] and the Gaunt and Breit potential energies in the quantum theory.[21]

Note: The Darwin potential energy was derived in 1920 by Charles Galton Darwin, grandson of the famous naturalist in 1920. The Gaunt and Breit potentials were derived by John Arthur Gaunt and Gregory Breit, respectively. The Dirac-Coulomb-Breit Hamiltonian, which consists of Dirac terms for the electrons plus Coulomb and Breit potential energies for electron-electron interactions, is a very popular special-relativistic quantum mechanical Hamiltonian used in atomic and molecular studies.

The second comment is that we can use the relationships between fields and potentials to show that $\epsilon_0(\boldsymbol{E}^2 + c^2\boldsymbol{B}^2)$ contains certain components that are proportional to the potentials Φ and $\boldsymbol{A}$, components that cannot be assigned to the electromagnetic field. The situation is even worse when there is no radiation, because in that case, the potentials depend exclusively on the matter variables and we cannot assign such components to a supposed radiation field external to the system of charges.

Third, the terms associated with the system of charges in the interpretations (4.4), (4.5), and (4.6) are counting the interactions twice. For example, we can easily verify that $\sum_i^N q_i\Phi$ gives twice the value of the Coulomb potential energy (3.4) for a system of charges, while terms like $\sum_i^N \boldsymbol{p}_i q_i \boldsymbol{A}/m_i c$ give twice the value that we would expect for the magnetic energy. This defect is a consequence of the terms corresponding to the charge-charge interaction being incorrectly included in $\epsilon_0(\boldsymbol{E}^2 + c^2\boldsymbol{B}^2)$, as we mentioned in the second comment above. Furthermore, the system identified as a system of free charges in (4.4) actually represents a system of magnetically interacting charges, because the energy of free charges is $\sum_i^N \sqrt{m_i^2c^4 + \boldsymbol{p}_i^2c^2}$.

The Hamiltonian formalism reveals another deficiency of the field model. Physicists like Griffiths insist that fields are physical systems in their own right, "*every bit as 'real' as atoms or baseballs*".[5] But, if the fields were really physical systems on an equal footing with matter, then the complete Hamiltonian for electromagnetism would consist of three terms: one for the system of charges, depending only on the matter variables; another term for the field, depending only on the field variables; and a third term for the interaction between charges and field, depending on field and matter variables. Formally,

$$H = H_{\mathrm{CHARGES}} + H_{\mathrm{INTERACTION}} + H_{\mathrm{FIELD}}. \tag{4.7}$$

However, the common interpretations (4.4), (4.5), and (4.6) show that this is not the case, suggesting that the fields are not really "*as atoms or baseballs*".

We will demonstrate in appendix A that the electromagnetic field is not a physical system in its own right, but rather this field can be understood as a residual component of the complete interaction between charges in the action-at-a-distance model.

The main **conclusions** of this chapter are the following:

- Some classical electrodynamics textbooks either avoid mentioning Lagrangians and Hamiltonians or give only incomplete and approximate forms for them. This is surprising considering that a Lagrangian or a Hamiltonian should completely describe the dynamics of a system of charges and the electromagnetic field.

- The Lagrangian and the Hamiltonian for a system of charges and field are divergent; that is $L = \infty$ and $H = \infty$. The Lagrangian and the Hamiltonian are formally infinite because they include nonphysical self-interacting field effects.

- There is no consensus among physicists on how to physically interpret each term in these Lagrangians and Hamiltonians. Furthermore, the usual interpretations found in the literature are inconsistent, because all the terms exclusively associated with the electromagnetic field depend on the matter variables and vice versa; for example, the term that is often identified as representing the energy stored in the field contains components that describe the potential energy of a system of charges.

- The usual interpretation of the electromagnetic field as a physical system on an equal footing with matter is untenable, because if this were the case, the complete Hamiltonian for a system of charges and electromagnetic field would have to consist of three terms: one for the matter, another for the field, and a third term for the interaction between matter and field.

—5

DELAYED INTERACTIONS

Often found in the physics literature is the claim that instantaneous interactions have been experimentally refuted in the laboratory. An example is the following quote, taken from the celebrated course[7] by Landau and Lifshitz: "*Experiment shows that instantaneous interactions do not exist in nature*". However, I will demonstrate in this chapter that those claims are not true.

We begin our analysis of the nature of electromagnetic interactions with the potential formulation of the Maxwell equations in the Lorenz gauge. Equations (2.11) and (2.12) are mathematically similar, allowing us to use the same method to solve both equations simultaneously. We will use the method[6,8–10] of the retarded Green functions to obtain the *retarded potentials*

$$\Phi_{\text{ret}}^{\text{Lor}}(\boldsymbol{r},t) = \frac{1}{4\pi\epsilon_0}\int \frac{\rho(\boldsymbol{r}',t^{\text{ret}})}{|\boldsymbol{r}-\boldsymbol{r}'|}\mathrm{d}\boldsymbol{r}' \tag{5.1}$$

and

$$\boldsymbol{A}_{\text{ret}}^{\text{Lor}}(\boldsymbol{r},t) = \frac{1}{4\pi\epsilon_0 c^2}\int \frac{\boldsymbol{J}(\boldsymbol{r}',t^{\text{ret}})}{|\boldsymbol{r}-\boldsymbol{r}'|}\mathrm{d}\boldsymbol{r}'. \tag{5.2}$$

As before, we are using the superscript 'Lor' to indicate that we are working on this gauge. The subscript 'ret' means that the value of the potentials at time t depends on the values of the densities and currents at the earlier time $t^{\mathrm{ret}} = (t - |\boldsymbol{r} - \boldsymbol{r}'|/c)$.

The pair of solutions (5.1) and (5.2) are often used to interpret the experiments Landau and Lifshitz refer to. These solutions to the Maxwell equations are taken as proof that electromagnetic interactions between charges are not instantaneous because there is a delay $(t - t^{\mathrm{ret}})$ between cause and effect. In the words of Landau and Lifshitz:[7]

> *"In actuality, if any change takes place in one of the interacting bodies, it will influence the other bodies only after the lapse of a certain interval of time. It is only after this time interval that processes caused by the initial change begin to take place in the second body."*

This delay is commonly interpreted in the physics literature as the time required for electromagnetic signals (photons) to travel from source charges to test charges:[8]

> *"The quotient $|\boldsymbol{r} - \boldsymbol{r}'|/c$ is just the time needed by a wave traveling at the speed of light to propagate from $\boldsymbol{r}'$ to $\boldsymbol{r}$. So, the principle of causality for perturbations propagating into the future is taken into account by retarded potentials."*

The propagation of electromagnetic signals mentioned above by Greiner[8] is illustrated in figure 5.1 in terms of past and future light cones. The physical situation described by the retarded potentials corresponds to the lower cone.

But before continuing our discussion of solutions of the Maxwell equations, we would point out a very important error that can be found in the literature on classical electrodynamics. This error is related to the propagation of electromagnetic signals.

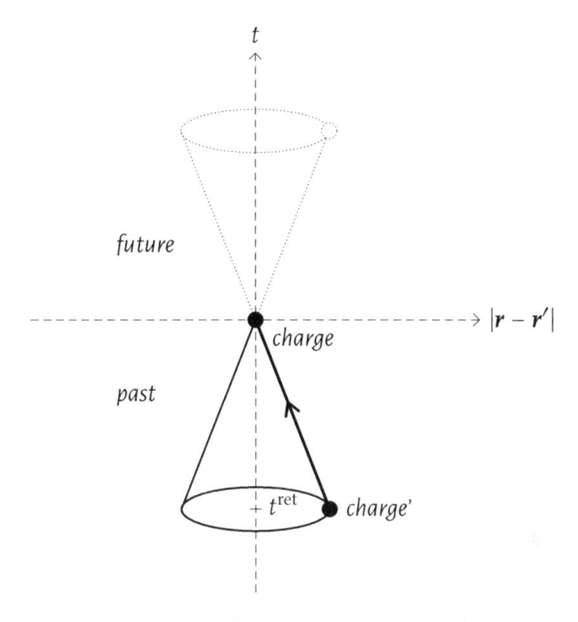

Figure 5.1: Light cone representation of the propagation of an electromagnetic wave from one charge to another. The lower cone represents the arrival in the present of a wave emitted by another charge in the past, a situation that corresponds to the retarded potentials. The physical interpretation of the upper cone will be discussed later in this same chapter.

Consider the following excerpt from Landau and Lifshitz's course on field theory:[7]

> "*From the principle of relativity it follows in particular that the velocity of propagation of interactions is the same in all inertial systems of reference. Thus the velocity of propagation of interactions is a universal constant. This constant velocity (as we shall show later) is also the velocity of light in empty space. The velocity of light is usually designated by the letter c, and its numerical value is* $c = 2.998 \times 10^{10}$ *cm/sec.*"

As explained in general physics courses, a velocity is a *vector quantity* because it requires of a vector for its complete description.

However, c in the excerpt above is a *scalar quantity*, since it is given only by a numerical value and a unit. As mentioned in chapter 2, this scalar quantity c is the speed of light in the vacuum and its modern SI value is exactly equal to 299,792,458 meters per second. c is not a velocity.

Note: Initially, the speed of light was a measured quantity and, consequently, its value was affected by both experimental errors and measurement uncertainties. However, since 1983 c has been defined to have an exact value, with all uncertainties and errors absorbed within the quantities of position and time. The new definition was proposed by international agreement at the 17th *General Conference on Weights and Measures* (abbreviated to CGPM for the French *Conférence générale des poids et mesures*).

The same error is reproduced[5,6,8] in many other standard textbooks, with their respective authors referring to c as "*the velocity of light*", when c is a speed.

The aforementioned authors not only confuse a scalar with a vector, but they also misinterpret the physical nature of electromagnetic signals and their role in describing interactions in the field model. It can be shown that photons cannot have a velocity, since they are not localizable particles. Even if we assign an initial velocity to photons, this velocity becomes unobservable after a short period of time. Photons can only have speed and this speed is c in the vacuum, as we show in appendix B.

After this brief deviation from the main topic, we can continue our discussion of the solutions to the Maxwell equations in the potential formulation. If we choose the Coulomb gauge, then the equations are (2.14) and (2.15). We can rewrite the second equation as

$$\nabla^2 \mathbf{A}^{\mathrm{Coul}} = -\mu_0 \boldsymbol{J}^{\mathrm{T}}, \tag{5.3}$$

with the help of what Maxwell[22] called the "*true current*" $\boldsymbol{J}^{\mathrm{T}}$. Of course, the superscript 'Coul' indicates that we are working on this gauge.

Maxwell took the true electromagnetic current $\boldsymbol{J}^{\mathrm{T}}$ to be the sum of $\boldsymbol{J}$ and the displacement current

$$\boldsymbol{J}^{\mathrm{T}} = \boldsymbol{J} \underbrace{- \frac{1}{\mu_0 c^2}\frac{\partial^2 \boldsymbol{A}^{\mathrm{Coul}}}{\partial t^2} - \frac{1}{\mu_0 c^2}\frac{\partial}{\partial t}(\nabla \Phi^{\mathrm{Coul}})}_{\text{DISPLACEMENT CURRENT}}. \tag{5.4}$$

Equation (5.3) is mathematically similar to (2.14), which allows us to use the method of the instantaneous Green functions to solve both equations simultaneously, obtaining the *instantaneous potentials*

$$\Phi^{\mathrm{Coul}}_{\mathrm{inst}}(\boldsymbol{r}, t) = \frac{1}{4\pi\epsilon_0}\int \frac{\rho(\boldsymbol{r}', t)}{|\boldsymbol{r} - \boldsymbol{r}'|}\mathrm{d}\boldsymbol{r}' \tag{5.5}$$

and

$$\boldsymbol{A}^{\mathrm{Coul}}_{\mathrm{inst}}(\boldsymbol{r}, t) = \frac{1}{4\pi\epsilon_0 c^2}\int \frac{\boldsymbol{J}^{\mathrm{T}}(\boldsymbol{r}', t)}{|\boldsymbol{r} - \boldsymbol{r}'|}\mathrm{d}\boldsymbol{r}', \tag{5.6}$$

where the subscript 'inst' means that the value of the potentials at time t depends on the values of the densities and currents at the same time t.

Using the Coulomb gauge, we have found that the values of the scalar and vector potentials at time t depend on the values of the density and the *true current* at the same time t. In this case there is no delay between cause and effect, contrary to the common wisdom we quoted on page 46.

Therefore, the following fundamental questions arise: are electromagnetic interactions instantaneous or delayed? Are $\Phi^{\mathrm{Lor}}_{\mathrm{ret}}$ and $\boldsymbol{A}^{\mathrm{Lor}}_{\mathrm{ret}}$ the correct potentials or are $\Phi^{\mathrm{Coul}}_{\mathrm{inst}}$ and $\boldsymbol{A}^{\mathrm{Coul}}_{\mathrm{inst}}$? Or maybe both sets of potentials are correct and describe the same physics?

In 1894, Poincaré noted that:[22]

> *"In calculating [$\boldsymbol{A}^{\text{Coul}}_{\text{inst}}$] Maxwell takes into account the currents of conduction and those of displacement; and he supposes that the attraction takes place according to Newton's law, i.e. instantaneously. But in calculating [$\boldsymbol{A}^{\text{Lor}}_{\text{ret}}$] on the contrary we take account only of conduction currents and we suppose the attraction is propagated with the velocity of light. It is a matter of indifference whether we make this hypothesis of a propagation in time and consider only the induction due to conduction currents, or whether like Maxwell, we retain the old law of instantaneous induction and consider both conduction and the displacement currents."*

The lesson we learn from Maxwell and Poincaré is the existence of a dual interpretation of the electromagnetic interactions. We can describe all the observed electromagnetic phenomena in terms of instantaneous interactions for the sources ($\rho, \boldsymbol{J}^{\text{T}}$) or in terms of retarded interactions with sources ($\rho, \boldsymbol{J}$). Both the retarded and the instantaneous descriptions of the electromagnetic interactions are formally correct, because each is a consequence of the Maxwell equations.

I have presented what I call a symmetrical view of the potentials for the two main gauge options: the retarded potentials $\Phi^{\text{Lor}}_{\text{ret}}$ and $\boldsymbol{A}^{\text{Lor}}_{\text{ret}}$ in the Lorenz gauge, and the instantaneous potentials $\Phi^{\text{Coul}}_{\text{inst}}$ and $\boldsymbol{A}^{\text{Coul}}_{\text{inst}}$ in the Coulomb gauge. This presentation is consistent with the historical development of electromagnetic theory and with recent literature on the foundations of electromagnetism, but it is inconsistent with the view presented to students in modern textbooks.

Most physicists today still use the retarded potentials in the Lorenz gauge, but they offer an asymmetric view of the potentials in the Coulomb gauge. In this view, $\Phi^{\text{Coul}}_{\text{inst}}$ is kept intact, but equation (2.15) for the vector potential is rearranged as

$$\nabla^2 \boldsymbol{A}^{\text{Coul}} - \frac{1}{c^2}\frac{\partial^2 \boldsymbol{A}^{\text{Coul}}}{\partial t^2} = -\mu_0 \boldsymbol{J}^{\perp}, \tag{5.7}$$

where the "*transverse current*" $\boldsymbol{J}^{\perp}$ is defined by

$$\boldsymbol{J}^{\perp} = \boldsymbol{J} - \frac{1}{\mu_0 c^2}\frac{\partial}{\partial t}(\nabla \Phi^{\text{Coul}}), \tag{5.8}$$

and then solved with the help of the retarded Green functions,

$$\boldsymbol{A}^{\text{Coul}}_{\text{ret}}(\boldsymbol{r}, t) = \frac{1}{4\pi\epsilon_0 c^2}\int \frac{\boldsymbol{J}^{\perp}(\boldsymbol{r}', t^{\text{ret}})}{|\boldsymbol{r} - \boldsymbol{r}'|}\mathrm{d}\boldsymbol{r}'. \tag{5.9}$$

This popular choice of asymmetric potentials, with an instantaneous scalar potential $\Phi^{\text{Coul}}_{\text{inst}}$ and a retarded vector potential $\boldsymbol{A}^{\text{Coul}}_{\text{ret}}$ in terms of a transverse current, is the reason why the Coulomb gauge is often also called the "*transverse gauge*".[6]

In summary, we have obtained three types of solutions of the Maxwell equations in their potential form:

- The pair of retarded potentials $\Phi^{\text{Lor}}_{\text{ret}}$ and $\boldsymbol{A}^{\text{Lor}}_{\text{ret}}$ in the Lorenz gauge. Equations (5.1) and (5.2).
- The pair of instantaneous potentials $\Phi^{\text{Coul}}_{\text{inst}}$ and $\boldsymbol{A}^{\text{Coul}}_{\text{inst}}$ in the Coulomb gauge. Equations (5.5) and (5.6).
- The asymmetric pair composed of an instantaneous scalar potential $\Phi^{\text{Coul}}_{\text{inst}}$ and a retarded transverse potential $\boldsymbol{A}^{\text{Coul}}_{\text{ret}}$. Equations (5.5) and (5.9).

All three types of solutions are physically valid and can be used to solve real-world problems. However, the instantaneous potentials in the Coulomb gauge are not frequently found in modern textbooks because the physical nature of the Maxwell displacement current in (5.4) has been controversial since the concept was first introduced in 1862. A survey of recent discussions can be found in the papers by Roche, Jefimenko, Jackson, and Heras.[23–29]

Arguments against the displacement current range from weak comments to obvious mistakes. I will not present here all the arguments found in the cited literature, but I will provide, for illustration purposes, some of those arguments against the displacement current.

To begin, consider the following statement:[24] "*for calculating **B** in time-dependent systems by means of equation [...], we need to know **B** in the first place!*". This is not a valid criticism of the displacement current because Jefimenko omits that we can apply an iteration procedure to obtain the magnetic field.

Note: As early as 2000 BC, the Babylonians used iterative procedures to find square roots. Iteration procedures are routinely performed in quantum mechanics, for example, to solve the Lippmann-Schwinger equation

$$\Psi = \Psi_0 + \frac{1}{E - \hat{H}_0 + \mathrm{i}\epsilon}\hat{U}\Psi.$$

Do not worry now about what each symbol in this equations means, just pay attention to the fact that we need to introduce the solution Ψ on the right side of the equation before solving it to obtain Ψ from the left side. What can we do? We can first replace Ψ with Ψ_0 on the right side of the equation, where Ψ_0 is the solution when the system is free ($\hat{U} = 0$). We solve the equation with Ψ_0 on the right side and obtain an approximate solution Ψ_1 from the left side.

$$\Psi_1 = \Psi_0 + \frac{1}{E - \hat{H}_0 + \mathrm{i}\epsilon}\hat{U}\Psi_0.$$

This approximate solution Ψ_1 is then introduced on the right side in order to calculate an improved solution Ψ_2,

$$\Psi_2 = \Psi_0 + \frac{1}{E - \hat{H}_0 + \mathrm{i}\epsilon}\hat{U}\Psi_1.$$

which can be introduced again on the right side in the following iteration. The procedure is repeated until the difference between two consecutive solutions is less than the precision of the laboratory equipment, finally taking any of them as the solution Ψ.

Another argument against the displacement current is Roche's claim that "*the Coulomb gauge is not a physical gauge*".[25] Roche is not only ignoring the true lesson of gauge theory, that all gauges produce the same physical results, but he also omits that the Coulomb gauge is *the preferred choice* for special-relativistic quantum mechanical atomic and molecular computations. How could a nonphysical gauge produce highly accurate predictions that fully agree with the experimental findings confirmed in the laboratory? Roche maintains that we would differentiate between numerical and physical solutions to the Maxwell equations. He offers the counterexample of how we can calculate the magnetic intensity of a narrow solenoid using either current loops or using magnetic monopoles, even though the monopoles do not exist.[30] However, the relationship between loops and monopoles is not the kind of relationship that exists between the Lorenz and Coulomb gauges, as we will see in a moment.

Heras' analysis[29] shows that the displacement current is a true physical current, although not a local one. So why is Maxwell's discovery so overwhelmingly rejected in the usual electrodynamics literature? A careful reading of the criticism against the use of the displacement current reveals that its critics reject Maxwell's discovery because the use of this current in a vector potential like (5.6) allows an instantaneous description of electromagnetic phenomena. Roche expresses his concern in the following terms:[23]

> "*The Maxwell expression asserts that these fields are transmitted instantaneously to the observation point while the Lorentz expression implies that the transmission is retarded at the speed of light. It is hardly necessary to point out the very serious difficulties involved in the first interpretation. Most physicists already seem to accept the second, which is strongly supported by special relativity and by relativistic quantum electrodynamics.*"

Jefimenko[24] expresses himself in similar terms:

> *"Detailed investigations of causal relations in electromagnetic fields have shown [...] that the only viable way to compute electric and magnetic fields in terms of their causative sources is to use retarded field integrals (or retarded potentials)."*

We can also find misstatements about how retarded interactions are required for accuracy:[7]

> *"However, experiment shows that instantaneous interactions do not exist in nature. Thus a mechanics based on the assumption of instantaneous propagation of interactions contains within itself a certain inaccuracy."*

However, the instantaneous potentials (5.5) and (5.6) are an *exact solution* to the Maxwell equations and provide the same accuracy in describing the phenomena as the retarded potentials, Furthermore, the claim that experiments disproved instantaneous interactions is blatantly false, as mentioned at the beginning of this chapter.

It seems that Maxwell and Poincaré were aware of the difference between the action-at-a-distance and field models more than a century ago, since in the quote reproduced on page 50, Poincaré carefully uses the term "*propagation*" only for the model of interactions based on an intermediary electromagnetic field.

We can see that even recent Nobel Prize laureates like Landau misunderstand the action-at-a-distance model and believe that instantaneous interactions are not only inaccurate but disproved by experiment. This type of confusion can be removed from the physics literature by reviewing the deeper technical details of the field model and of the action-at-a-distance model. This is just what we are going to do next.

In field theory, electromagnetic interactions between charges are indirect and require the propagation of a signal through an intermediate medium as shown in diagram (3.2). According to the laws of special relativity, nothing can travel faster than light, and this inevitably introduces a delay between the time that the source charge emits an electromagnetic signal and the time that the test charge receives it, as shown in figure 5.1.

We can conclude that an instantaneous propagation of electromagnetic signals between charged particles defies the laws of the theory of special relativity. This line of reasoning is essentially correct, and Roche and Jefimenko would have every right to reject the instantaneous formulation of the Maxwell theory if this postulated faster-than-light signals. However, the action-at-a-distance model is completely different, since *nothing travels faster than light in the instantaneous formulation* of the Maxwell theory. As we explained in chapter 3, the charges interact directly as shown in diagram (3.1), *without sending or receiving electromagnetic signals*. There is no electromagnetic field in the action-at-distance model, only charges interacting directly without any intermediary. The absence of any propagating signal is what makes the instantaneous interactions fully causal and compatible with special relativity.[31]

Note: The difference between both models of interactions is perhaps best understood from a quantum perspective. In quantum field theory, light signals are made up of photons, and photons are in some way associated with excitations of the electromagnetic field. There is no field in the action-at-a-distance model, and therefore charges cannot emit or absorb photons to interact.

Let me emphasize this one more time because it is very important: since charges do not emit signals in the action-at-a-distance model, the instantaneous potentials (5.5) and (5.6) do not violate special relativity, the principle of causality, or any known experiment. The key here is that the predictions about the motion of charged particles obtained in the Coulomb gauge with the help of the "*true current*" $\boldsymbol{J}^{\mathrm{T}}$ are physically identical to the predictions obtained in the Lorenz gauge using only the ordinary current $\boldsymbol{J}$.

At this point we must recover Roche's statement that "*the Coulomb gauge is not a physical gauge*".[25] As Jackson mentions in a reply to him:[28]

> "*The choice of gauge is purely a matter of convenience. The Coulomb gauge is no more or less physical than any other. It is convenient for some problems, inconvenient for others.*"

and Jackson continues in the following terms:

> "*The fields are the reality. The Coulomb gauge has potentials with peculiar ('unphysical') relativistic properties, but the fields derived from them are the same as the fields derived from the potentials in any other gauge, causal and with finite speed of propagation.*"

This is precisely the type of misconception that we denounced on the previous pages when we studied the deeper technical details of the field and action-at-a-distance models. The Coulomb gauge yields potentials with 'unphysical' relativistic properties *only when one insists on interpreting those potentials in terms of the field model of interactions, with electromagnetic signals propagating from one charge to another*. The scalar potential in the Coulomb gauge, in general any instantaneous potential, is entirely physical when interpreted in terms of the action-at-a-distance model, as Maxwell and Poincaré did more than one century ago.

The interpretation of the instantaneous formulation of Maxwell's theory in terms of the action-at-a-distance model can also be found in the recent work of Heras:[29]

> "*Maxwell's equations in the form given by Eqs. [...] represent an instantaneous action-at-a-distance theory.*"

Heras' paper is very helpful in dispelling myths about the instantaneous formulation of electromagnetic interactions, but his discussion of issues regarding causality, signals propagation at the speed of light, and the physical role of the fields is not entirely complete. For example, Heras claims that "*both formulations are formally equivalent*", but an action-at-a-distance model of the electromagnetic interactions is actually equivalent to a field model *with renormalization corrections*, as we will see in appendix A,

$$\textbf{\textit{action at a distance}} = \textbf{\textit{field theory}} + \textbf{\textit{renormalization}}. \tag{5.10}$$

Heras, following Maxwell and Poincaré, associates the instantaneous formulation with the Coulomb gauge and the retarded formulation with the Lorenz gauge, but an instantaneous formulation of electromagnetic interactions is also available in the Lorenz gauge. Consider equations (2.11) and (2.12); we solved them using the values of the sources ρ and $\boldsymbol{J}$ at the early time t^{ret}. But if we move the double time-derivatives to the right side of the equations and introduce new sources Ω and $\boldsymbol{S}$ as in

$$\nabla^2 \Phi^{\mathrm{Lor}} = -\left[\frac{\rho}{\epsilon_0} - \frac{1}{c^2}\frac{\partial^2 \Phi^{\mathrm{Lor}}}{\partial t^2}\right] = -\frac{\Omega}{\epsilon_0} \tag{5.11}$$

and

$$\nabla^2 \boldsymbol{A}^{\mathrm{Lor}} = -\left[\mu_0 \boldsymbol{J} - \frac{1}{c^2}\frac{\partial^2 \boldsymbol{A}^{\mathrm{Lor}}}{\partial t^2}\right] = -\mu_0 \boldsymbol{S}, \tag{5.12}$$

then we can obtain the following pair of instantaneous potentials

$$\Phi^{\mathrm{Lor}}_{\mathrm{inst}}(\boldsymbol{r}, t) = \frac{1}{4\pi\epsilon_0}\int \frac{\Omega(\boldsymbol{r}', t)}{|\boldsymbol{r} - \boldsymbol{r}'|} \mathrm{d}\boldsymbol{r}' \tag{5.13}$$

and

$$\boldsymbol{A}^{\mathrm{Lor}}_{\mathrm{inst}}(\boldsymbol{r}, t) = \frac{1}{4\pi\epsilon_0 c^2}\int \frac{\boldsymbol{S}(\boldsymbol{r}', t)}{|\boldsymbol{r} - \boldsymbol{r}'|} \mathrm{d}\boldsymbol{r}'. \tag{5.14}$$

The value of the potentials at time t depends on the values of the sources Ω and $\boldsymbol{S}$ at the same time.

Without any doubt, also in the Lorenz gauge we can find a double description of the interactions, either in terms of retarded potentials (5.1) and (5.2) for the ordinary sources ρ and $\boldsymbol{J}$ or in terms of instantaneous potentials (5.13) and (5.14) for the alternative sources Ω and $\boldsymbol{S}$. The alternative sources are precisely what allows the existence of the Darwin Hamiltonian (and its associated Lagrangian) in the classical theory[20,21] and its quantum counterpart: the Dirac-Coulomb-Breit Hamiltonian.[21]

Some physicists might argue that Ω and $\boldsymbol{S}$ are not 'true' sources, since that they have been defined in (5.11) and (5.12) in terms of the sources ρ and $\boldsymbol{J}$, and the double partial derivatives $(\partial^2\Phi^{\mathrm{Lor}}/\partial t^2)$ and $(\partial^2\boldsymbol{A}^{\mathrm{Lor}}/\partial t^2)$. However, this recurrence relationship is a direct consequence of starting with equations (2.11) and (2.12). If instead we start with the equations

$$\nabla^2\Phi_{\mathrm{inst}}^{\mathrm{Lor}} = -\frac{\Omega}{\epsilon_0} \tag{5.15}$$

and

$$\nabla^2\boldsymbol{A}_{\mathrm{inst}}^{\mathrm{Lor}} = -\mu_0\boldsymbol{S}, \tag{5.16}$$

and we compare them with equations (2.11) and (2.12), then we obtain the identities

$$\rho = \left[\Omega + \frac{\epsilon_0}{c^2}\frac{\partial^2\Phi_{\mathrm{inst}}^{\mathrm{Lor}}}{\partial t^2}\right] \tag{5.17}$$

and

$$\boldsymbol{J} = \left[\boldsymbol{S} + \frac{1}{\mu_0 c^2}\frac{\partial^2\boldsymbol{A}_{\mathrm{inst}}^{\mathrm{Lor}}}{\partial t^2}\right]. \tag{5.18}$$

That is, we can obtain the retarded sources ρ and $\boldsymbol{J}$ from the instantaneous sources Ω and $\boldsymbol{S}$, and vice versa, since the descriptions are equivalent. So far, our discussion has been general and abstract, but we will see a practical example in chapter 8, where we will show that the Lorentz Lagrangian using ρ and $\boldsymbol{J}$ is equivalent to the Darwin Lagrangian using Ω and $\boldsymbol{S}$.

Note: The Darwin Lagrangian is equivalent to the Lorentz Lagrangian up to c^{-2} order of accuracy in a power series expansion $[1 + \mathcal{O}(c^{-2}) + \mathcal{O}(c^{-3}) + \cdots]$ in which the first term corresponds to the Newton-Coulomb description and the notation $\mathcal{O}(c^{-n})$ means that we are considering special-relativistic effects up to order n. Darwin worked in the c^{-2} approximation to avoid dealing with the difficulties associated to radiation in field theory, but it is possible to obtain a generalized instantaneous Lagrangian valid for arbitrary order.

The instantaneous potentials (5.13) and (5.14) in the Lorenz gauge are physically identical to the retarded potentials (5.1) and (5.2) in the same gauge because the following identities hold

$$\int \frac{\Omega(\boldsymbol{r}', t)}{|\boldsymbol{r} - \boldsymbol{r}'|} \mathrm{d}\boldsymbol{r}' = \int \frac{\rho(\boldsymbol{r}', t^{\mathrm{ret}})}{|\boldsymbol{r} - \boldsymbol{r}'|} \mathrm{d}\boldsymbol{r}' \tag{5.19}$$

and

$$\int \frac{\mathbf{S}(\boldsymbol{r}', t)}{|\boldsymbol{r} - \boldsymbol{r}'|} \mathrm{d}\boldsymbol{r}' = \int \frac{\boldsymbol{J}(\boldsymbol{r}', t^{\mathrm{ret}})}{|\boldsymbol{r} - \boldsymbol{r}'|} \mathrm{d}\boldsymbol{r}', \tag{5.20}$$

and we can go from the instantaneous to the retarded potentials and vice versa. This equivalence is used by Villeco to develop a "*relativistically correct instantaneous action-at-a-distance representation of field theories*".[32]

This dual representation of electromagnetic phenomena is not limited to the potentials; for any mechanical or electromagnetic function G evaluated at retarded times, there exists a complementary function K that gives the same value when evaluated at the present time

$$G(t^{\mathrm{ret}}) = K(t). \tag{5.21}$$

We are now going to continue studying the solutions of the electromagnetic potentials in the Lorenz gauge but changing the focus a bit. The title of this chapter is "*delayed interactions*" rather than retarded interactions because, although so far we have solved equations (2.11) and (2.12) using the standard method of the retarded Green functions, we could have used advanced Green functions, in which case we would obtain the following pair of alternative solutions

$$\Phi^{\mathrm{Lor}}_{\mathrm{adv}}(\boldsymbol{r}, t) = \frac{1}{4\pi\epsilon_0} \int \frac{\rho(\boldsymbol{r}', t^{\mathrm{adv}})}{|\boldsymbol{r} - \boldsymbol{r}'|} \mathrm{d}\boldsymbol{r}' \tag{5.22}$$

and

$$\boldsymbol{A}^{\mathrm{Lor}}_{\mathrm{adv}}(\boldsymbol{r}, t) = \frac{1}{4\pi\epsilon_0 c^2} \int \frac{\boldsymbol{J}(\boldsymbol{r}', t^{\mathrm{adv}})}{|\boldsymbol{r} - \boldsymbol{r}'|} \mathrm{d}\boldsymbol{r}'. \tag{5.23}$$

In this alternative set of potentials, the electric densities and currents are evaluated at the future time $t^{\mathrm{adv}} = (t + |\boldsymbol{r} - \boldsymbol{r}'|/c)$. What does all this mean?

Most physicists reject the advanced potentials (5.22) and (5.23) because these potentials require knowing the value of the charge densities and currents in the future (see the upper light cone in figure 5.1). Griffiths gives the following arguments to rule out these potentials:[5]

> *"Although the advanced potentials are entirely consistent with Maxwell's equations, they violate the most sacred tenet in all of physics: the principle of causality. They suggest that the potentials now depend on what the charge and the current distribution will be at some time in the future—the effect; in other words, precedes the cause. Although the advanced potentials are of some theoretical interest, they have no direct physical significance."*

Schwinger et al. express similar concerns:[9]

> *"If we pick the − sign, we obtain the advanced Green's function [...] which is nonzero when the signal is arriving at the observer before it is emitted by the source. Since the latter is not in accordance with elementary ideas of causality, we adopt the retarded Green's function as the solution which satisfies the correct time boundary condition. (Actually, both retarded and advanced Green's functions are useful in physics.)"*

We can also find similar statements in the rest of the standard literature. However, a rejection of the advanced potentials based on an alleged violation of causality is easily refuted using the instantaneous representation of the advanced potentials, because in this case "*only the present-time state of the source particle motion is involved*"[32] and no effect precedes the cause.

Furthermore, when the aforementioned authors claim that the advanced potentials are useful in physics, what they really mean is that the advanced potentials are used to *correct fundamental difficulties with equations of motion that use only the retarded potentials*. We will ignore for now the technical details of these difficulties, which include the violation of the law of conservation of energy, but in chapter 7 we will discuss the role that the advanced potentials play in obtaining radiation reaction forces.

Now we should ask, how could nonphysical potentials be needed to correct physical potentials? If it were true that only the retarded potentials have "*direct physical significance*", then why are the advanced potentials necessary to remove the nonphysical predictions made by the retarded ones? Do not expect the mainstream literature to answer these questions, as it simply ignores them, but it seems evident that the advanced potentials cannot be dismissed lightly.

We now turn our attention to whether the advanced potentials "*violate the most sacred tenet in all of physics*" or whether this is just another misconception. The instantaneous representation of the advanced potentials shows that causality is not violated, but we can better understand the physical interpretation of these potentials if we analyze their causality from the point of view of the ordinary delayed representation.

The causality of the retarded potentials is generally interpreted according to the diagram shown in figure 5.1. The past light cone represents an electromagnetic signal that travels from the charge in the past to the charge in the present. Similarly, the advanced potentials are usually interpreted in terms of an electromagnetic signal traveling backward in time, from the charge in the future to the charge in the present, as shown in figure 5.2. However, there is no physical justification for this latter interpretation except if the goal is to consider past and future light cones in a time-symmetric way. We can offer an alternative physical interpretation of the advanced potentials.

We begin by considering how the Newton law of action and reaction applied to Coulomb interactions implies that if one charge exerts an interaction on a second charge

$$\textit{charge'} \longrightarrow \textit{charge}, \tag{5.24}$$

then the second charge exerts an interaction on the first charge

$$\textit{charge'} \longleftarrow \textit{charge}. \tag{5.25}$$

Contrary to what you may have been told, this law is also valid in a special-relativistic context and provides a basis for the action-at-a-distance model of interactions. If we use equivalences (5.19) and (5.20) to convert from the delayed to the instantaneous formulation, we find that the special-relativistic law of action and reaction is violated, as shown in figure 5.2, by the traditional interpretation of the advanced potentials because only the diagram (5.24) is satisfied under the causality requirement of causes preceding their effects.

However, if the advanced potential does not represent an electromagnetic wave traveling backward in time, but rather a wave emitted by the charge in the present and traveling towards the

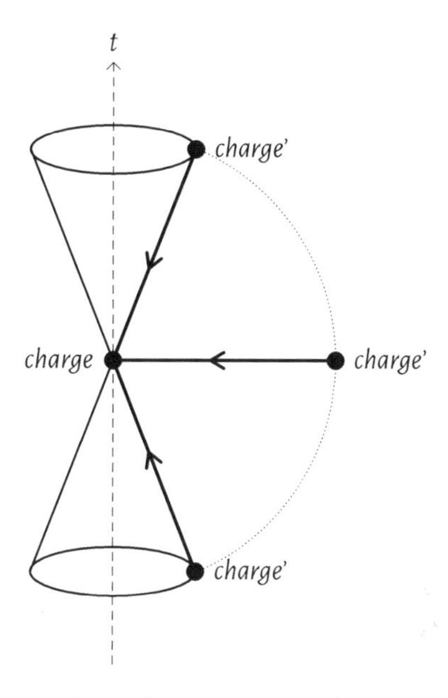

Figure 5.2: Ordinary interpretation of causality in the delayed formulation of electromagnetic interactions. If the advanced potential corresponds to the action of the charge in the future on the charge in the present time, then the special-relativistic law of action and reaction is violated.

future position of the second charge (that is, if the advanced potential is like the retarded potential but with a reversed role for the charges), then both diagrams (5.24) and (5.25) are included in the delayed potentials and a special-relativistic version of the law of action and reaction holds, as shown in figure 5.3.

We have suggested diagrammatically a new interpretation of the advanced potentials, but we can demonstrate it formally. The delayed potentials were obtained using the standard method of the Green functions. If we use the notation G^{ret} for the retarded Green function and G^{adv} for the advanced one, then we find by direct inspection that[10]

$$G^{\text{ret}}(\boldsymbol{r}, t; \boldsymbol{r}', t') = G^{\text{adv}}(\boldsymbol{r}', t'; \boldsymbol{r}, t), \tag{5.26}$$

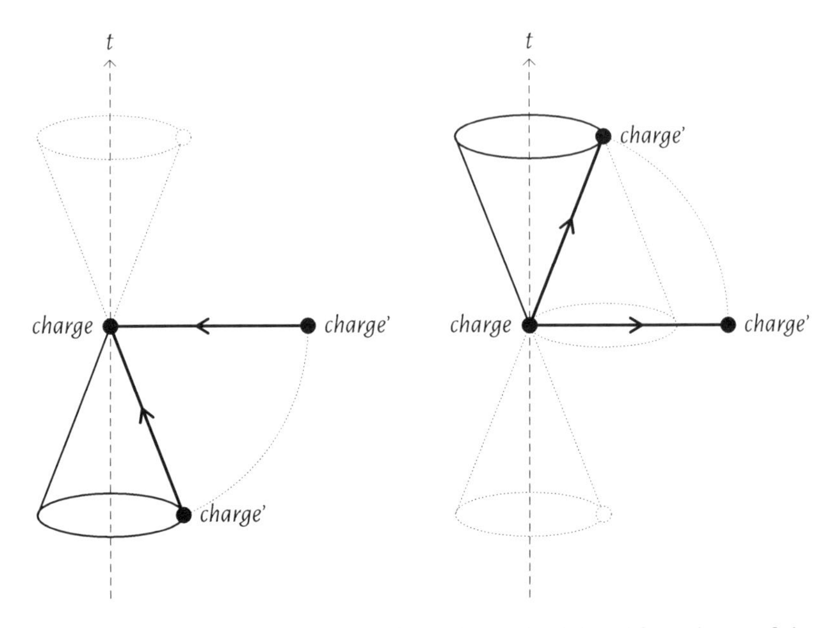

Figure 5.3: Alternative interpretation of causality in the delayed formulation of electromagnetic interactions. The advanced potential represents an electromagnetic wave emitted by the charge in the present time towards the future position of the second charge and the special-relativistic law of action and reaction is fulfilled.

which means that the advanced potential is equivalent to the retarded potential when the roles of the charges are exchanged. For example, using the advanced potential for (5.24) is equivalent to using the retarded potential for (5.25). This is exactly the case that we showed in figure 5.3.

Note: Our previous discussion has been completely classical, but we know that electromagnetic signals are associated with photons in quantum field theory. The Stueckelberg-Feynman interpretation of antiparticles states that a particle traveling backward in time is equivalent to an antiparticle traveling forward in time. Particles and antiparticles are distinguished by the sign of their electric charge. Photons have no charge, which means that the photon is the same particle as the antiphoton. So a photon traveling backward in time can be reinterpreted as a photon traveling forward in time.

The main **conclusions** of this chapter are the following:

- The claim that instantaneous interactions have been disproved in the laboratory is false. Different potentials can generate the same fields and there is a dual interpretation of electromagnetic phenomena in terms of instantaneous or retarded potentials depending on the nature of the sources.
- Arguments against the Maxwell displacement current range from weak comments to obvious mistakes. An example of an obvious mistake is the statement that "*the Coulomb gauge is not a physical gauge*".
- Instantaneous potentials are completely causal and compatible with both special relativity and all known experiments, because no signal is traveling faster than light. The instantaneous formulation interpreted in the context of an action-at-a-distance model of interactions is equivalent to the usual field model of interactions when we add renormalization corrections. The instantaneous formulation is not exclusive to the Coulomb gauge and the retarded formulation is not exclusive to the Lorenz gauge. The equivalence between instantaneous and retarded potentials can be used to obtain a "*relativistically correct instantaneous action-at-a-distance representation of field theories*".[32] The argument that the advanced potentials violate the principle of causality is also invalid.
- Advanced potentials do not represent an electromagnetic signal traveling backward in time, from the charge in the future to the charge in the present, but rather a wave emitted by the charge in the present and traveling towards the future position of the second charge.

—6

ELECTROMAGNETIC FORCES

According to standard electromagnetic literature, the Lorentz force (2.16) is the force acting on a point charge q moving with velocity $\boldsymbol{v}$ in the presence of an electromagnetic field, but this common statement found in textbooks[5,6,8–10] is misleading for at least three basic reasons.

First reason: there is no single definition of force in physics and stating, as Jackson does, that the Lorentz force is "*the total electromagnetic force on a charged particle*"[6] is not acceptable. In this chapter we will discuss three different concepts of force and their corresponding equations of motion.

The second reason is that those textbooks equate the Lorentz force to the rate of change of the supposed momentum of the particle, but this is not the force that should be used in a Hamiltonian dynamics to evaluate the change in the phase space state $(\boldsymbol{r}, \boldsymbol{p})$. As we will immediately demonstrate, we cannot evaluate the changes in the position and momentum of a charged particle using the Lorentz force.

Momentum is defined as $\boldsymbol{p} = m\boldsymbol{v}$ in Newtonian dynamics, with $\boldsymbol{v} = (\mathrm{d}\boldsymbol{r}/\mathrm{d}t)$ being the velocity of the particle. However, this simple definition has to be rejected in the case of very high velocities and in the presence of electromagnetic interactions. To obtain the relation between velocity and momentum we have to use the first of the Hamilton equations of motion

$$\frac{\mathrm{d}\boldsymbol{r}}{\mathrm{d}t} = \left(\frac{\partial H}{\partial \boldsymbol{p}}\right)_{\boldsymbol{r}}. \tag{6.1}$$

Ignoring the 'free field' term in (4.6), the Hamiltonian for a single charged particle is

$$H = \sqrt{m^2c^4 + (\boldsymbol{p} - q\boldsymbol{A})^2c^2} + q\Phi, \tag{6.2}$$

and the corresponding first equation of motion is

$$\frac{\mathrm{d}\boldsymbol{r}}{\mathrm{d}t} = \frac{(\boldsymbol{p} - q\boldsymbol{A})}{m\gamma}, \tag{6.3}$$

where γ is the time-dilation factor defined by

$$\gamma = \frac{\sqrt{m^2c^4 + (\boldsymbol{p} - q\boldsymbol{A})^2c^2}}{mc^2}. \tag{6.4}$$

This result confirms that the simple Newtonian relationship between momentum and velocity is no longer valid since the momentum of the charged particle is now given by $\boldsymbol{p} = m\gamma\boldsymbol{v} + q\boldsymbol{A}$.

Using the traditional definition of force as the rate of change of the momentum $(\mathrm{d}\boldsymbol{p}/\mathrm{d}t)$, we can obtain an expression for the force from the second of the Hamilton equations

$$\frac{\mathrm{d}\boldsymbol{p}}{\mathrm{d}t} = -\left(\frac{\partial H}{\partial \boldsymbol{r}}\right)_{\boldsymbol{p}}, \tag{6.5}$$

which yields

$$\frac{\mathrm{d}\boldsymbol{p}}{\mathrm{d}t} = q[\nabla_{\boldsymbol{A}}(\boldsymbol{v}\cdot\boldsymbol{A}) - \nabla\Phi], \tag{6.6}$$

where the velocity is given by (6.3); that is, the velocity is a phase-space quantity $\boldsymbol{v} = \boldsymbol{v}(\boldsymbol{r}, \boldsymbol{p})$. Moreover, we are using the Feynman notation in which the subscript means that the gradient operates only on the vector potential $\boldsymbol{A}$.

In conclusion, the force we have to use in Hamiltonian dynamics is not the Lorentz force $\boldsymbol{F}^{\mathrm{Lor}}$ but

$$\boldsymbol{F}^{\mathrm{Ham}} = q[\nabla_{\boldsymbol{A}}(\boldsymbol{v} \cdot \boldsymbol{A}) - \nabla\Phi]. \tag{6.7}$$

In an exercise of originality, we call to (6.7) the Hamiltonian force. We would come to the same conclusion if we had used Lagrangian dynamics instead of Hamiltonian dynamics, because the Lorentz force is not the force that we have to use with the Lagrange equation of motion for a charged particle

$$\frac{\mathrm{d}}{\mathrm{d}t}\left(\frac{\partial L}{\partial \boldsymbol{v}}\right)_{\boldsymbol{r}} = \left(\frac{\partial L}{\partial \boldsymbol{r}}\right)_{\boldsymbol{v}}. \tag{6.8}$$

The only difference between the force derived by the Lagrangian method and the one derived from the Hamiltonian would be that the first force is defined in a position-velocity space $(\boldsymbol{r}, \boldsymbol{v})$, while the second is in the phase space $(\boldsymbol{r}, \boldsymbol{p})$.

To better understand the concept of force, let us now look at the relationship between the Hamiltonian force (6.7) and the Lorentz force (2.16). If we multiply both sides of (6.3) by $m\gamma$ and take the time derivative we find

$$\frac{\mathrm{d}(m\gamma\boldsymbol{v})}{\mathrm{d}t} = \frac{\mathrm{d}\boldsymbol{p}}{\mathrm{d}t} - q\frac{\mathrm{d}\boldsymbol{A}}{\mathrm{d}t} = \boldsymbol{F}^{\mathrm{Ham}} - q\frac{\mathrm{d}\boldsymbol{A}}{\mathrm{d}t}. \tag{6.9}$$

Using mathematical identities such as $\mathrm{d}/\mathrm{d}t = \partial/\partial t + \boldsymbol{v}\cdot\nabla$ and $\nabla_{\boldsymbol{A}}(\boldsymbol{v} \cdot \boldsymbol{A}) = \boldsymbol{v} \times (\nabla \times \boldsymbol{A}) + (\boldsymbol{v} \cdot \nabla)\boldsymbol{A}$, we obtain

$$\frac{\mathrm{d}(m\gamma\boldsymbol{v})}{\mathrm{d}t} = q\left(\boldsymbol{v} \times (\nabla \times \boldsymbol{A}) - \nabla\Phi - \frac{\partial \boldsymbol{A}}{\partial t}\right). \tag{6.10}$$

Finally, using the definitions (2.6) and (2.7) for the electromagnetic field, we can show that the right hand side of the above equation of motion is just the Lorentz force

$$\frac{\mathrm{d}(m\gamma\boldsymbol{v})}{\mathrm{d}t} = q\left(\boldsymbol{E} + \boldsymbol{v} \times \boldsymbol{B}\right). \tag{6.11}$$

The relationship between the Hamiltonian and Lorentz forces is therefore

$$\boldsymbol{F}^{\mathrm{Ham}} = \boldsymbol{F}^{\mathrm{Lor}} + q\frac{\mathrm{d}\boldsymbol{A}}{\mathrm{d}t}. \tag{6.12}$$

Now we can verify that the Lorentz force does not give the change in the momentum $\boldsymbol{p}$ of the charged particle, but rather the change in the quantity $m\gamma\boldsymbol{v}$. Note that the Hamiltonian force and the Lorentz force coincide when $\boldsymbol{A} = \boldsymbol{0}$ because in this case $\boldsymbol{p} = m\gamma\boldsymbol{v}$.

But then, how do physicists such as Jackson and Griffiths interpret equation (6.11)? They redefine the momentum of the charged particle to be just $m\gamma\boldsymbol{v}$, ignoring the $q\boldsymbol{A}$ term. That is, they do $\boldsymbol{p} = \boldsymbol{\pi} + q\boldsymbol{A}$ and take $\boldsymbol{\pi} = m\gamma\boldsymbol{v}$ as the momentum of the charged particle. However, their identification of $\boldsymbol{\pi}$ with the momentum for the charge is inconsistent. Let us see why.

Using $\boldsymbol{\pi} = (\boldsymbol{p} - q\boldsymbol{A})$, the Hamiltonian (6.2) can be written as

$$H = \sqrt{m^2c^4 + \boldsymbol{\pi}^2c^2} + q\Phi, \tag{6.13}$$

and from the first of the Hamilton equations of motion

$$\frac{\mathrm{d}\boldsymbol{r}}{\mathrm{d}t} = \frac{\boldsymbol{\pi}}{m\gamma}, \tag{6.14}$$

we obtain the expected relationship $\boldsymbol{\pi} = m\gamma\boldsymbol{v}$. However, from the second of the Hamilton equations we could not obtain the equation of motion (6.11). We obtain instead the alternative (*invalid*) equation

$$\frac{\mathrm{d}\boldsymbol{\pi}}{\mathrm{d}t} = \frac{\mathrm{d}(m\gamma\boldsymbol{v})}{\mathrm{d}t} = -q\nabla\Phi, \tag{6.15}$$

and this equation agrees with the Lorentz equation (6.11) only when $\boldsymbol{A} = \mathbf{0}$. That is, we cannot obtain the equation of motion for the general case if we use the Hamiltonian (6.13) with $\boldsymbol{\pi}$ taken as the momentum of the charged particle. So what do Griffiths, Jackson, Greiner, and the rest of the physicists do to solve this difficulty?

The standard practice is to keep the Hamiltonian (6.2) for the charged particle but identifying its momentum only with the $(m\gamma\boldsymbol{v})$ term. However, this yields an inconsistent Hamiltonian dynamics, where $(\boldsymbol{r}, \boldsymbol{p})$ is no longer a phase space since $\boldsymbol{r}$ is the position of the charged particle, but $\boldsymbol{p}$ would not be its momentum. This inconsistency is directly related to the difficulties to physically interpret each term in the complete Hamiltonian (2.18) for the system of charges and electromagnetic field.

If we want to develop a consistent Hamiltonian dynamics, then we must admit that only taking $\boldsymbol{p} = (m\gamma\boldsymbol{v} + q\boldsymbol{A})$ as the momentum of the particle we can obtain the equations of motion for the phase space structure $(\boldsymbol{r}, \boldsymbol{p})$ for a single particle. The Lorentz equation of motion (6.11) is of course valid, because it can be derived from the second Hamilton equation, but the left side of the Lorentz equation is not the rate of change of the particle's momentum, because this is given by $(\mathrm{d}\boldsymbol{p}/\mathrm{d}t)$ and not by $(\mathrm{d}\boldsymbol{\pi}/\mathrm{d}t)$.

So far we have presented two different concepts of electromagnetic force: $\boldsymbol{F}^{\mathrm{Ham}}$ and $\boldsymbol{F}^{\mathrm{Lor}}$. We can obtain a third concept if we further expand the time derivative of $(m\gamma\boldsymbol{v})$ as

$$\frac{\mathrm{d}(m\gamma\boldsymbol{v})}{\mathrm{d}t} = m\frac{\mathrm{d}\gamma}{\mathrm{d}t}\boldsymbol{v} + m\gamma\frac{\mathrm{d}\boldsymbol{v}}{\mathrm{d}t}, \tag{6.16}$$

use the expression for the Lorentz force on the left-hand side of the above equation, divide both sides by γ, and note that $(\mathrm{d}\boldsymbol{v}/\mathrm{d}t)$ is the acceleration $\boldsymbol{a}$ of the charged particle.

The final result is the following equation of motion

$$m\boldsymbol{a} = \frac{1}{\gamma}\left[q(\boldsymbol{E} + \boldsymbol{v} \times \boldsymbol{B}) - m\frac{\mathrm{d}\gamma}{\mathrm{d}t}\boldsymbol{v}\right], \tag{6.17}$$

and the right-hand side of this new equation can be interpreted as a force in the Newtonian sense of the product of mass and acceleration ($m\boldsymbol{a} = \boldsymbol{F}^{\mathrm{Newt}}$). In a display of originality, we call this new force the Newtonian force.

Thus, we have found three definitions of force and their corresponding equations of motion

$$\frac{\mathrm{d}\boldsymbol{p}}{\mathrm{d}t} = \boldsymbol{F}^{\mathrm{Ham}}, \tag{6.18}$$

$$\frac{\mathrm{d}(m\gamma\boldsymbol{v})}{\mathrm{d}t} = \boldsymbol{F}^{\mathrm{Lor}}, \tag{6.19}$$

and

$$m\boldsymbol{a} = \boldsymbol{F}^{\mathrm{Newt}}. \tag{6.20}$$

All three are valid equations of motion and by direct inspection we can obtain the following relationships between the forces

$$\boldsymbol{F}^{\mathrm{Ham}} = \boldsymbol{F}^{\mathrm{Lor}} + q\frac{\mathrm{d}\boldsymbol{A}}{\mathrm{d}t} = \gamma\boldsymbol{F}^{\mathrm{Newt}} + m\frac{\mathrm{d}\gamma}{\mathrm{d}t}\boldsymbol{v} + q\frac{\mathrm{d}\boldsymbol{A}}{\mathrm{d}t}. \tag{6.21}$$

When the speeds of the charges are very small compared to the speed of light, the time-dilation factor is negligible ($\gamma \approx 1$) and its time derivative vanishes. In this case also $\boldsymbol{A} = \boldsymbol{0}$ and the three forces coincide $\boldsymbol{F}^{\mathrm{Ham}} = \boldsymbol{F}^{\mathrm{Lor}} = \boldsymbol{F}^{\mathrm{Newt}}$.

The misleading equation $m\boldsymbol{a} = \boldsymbol{F}^{\mathrm{Lor}}$ is very popular in elementary texts. In fact, Griffiths applies this equation to the example 5.2 in his textbook[5] to calculate the trajectory of a charged particle in "*cycloid motion*", and we can find this equation also in the rest of the electrodynamics literature (see, for example, equation 23.63

in the textbook[8] by Greiner). However, the proper equation of motion for the Lorentz force is (6.19). Wald emphasizes this fact:[10]

> *"This differs from the usual nonrelativistic equation of motion for a charged particle found in elementary texts only in that on the left side of this equation,* **v** *is replaced by* γ**v***."*

The reason why $m\boldsymbol{a} = \boldsymbol{F}^{\text{Lor}}$ is found in those texts is because their authors are considering experimental setups where $\boldsymbol{F}^{\text{Lor}} \approx \boldsymbol{F}^{\text{Newt}}$.

At the beginning of this chapter we mentioned the existence of at least three reasons why the usual textbook identification of the Lorentz force with "*the force*" acting on charged particles is misleading. The third reason is that $\boldsymbol{F}^{\text{Ham}}$, $\boldsymbol{F}^{\text{Lor}}$, and $\boldsymbol{F}^{\text{Newt}}$ do not describe the electromagnetic force of the charge on itself. This interaction is sometimes called the self-action and is a force predicted by field theory: "*The field concept naturally leads to self-action*".[9]

In the next chapter, electromagnetic self-forces and the modification they introduce in the equations of motion will be analyzed. In the meantime, let me emphasize that all the equations, definitions, and relationships in this chapter that involve the use of the electromagnetic potentials Φ and **A** make sense (they are physically meaningful and mathematically rigorous) *if and only if* those potentials are replaced by the so-called external potentials Φ^{ext} and $\boldsymbol{A}^{\text{ext}}$.

That is, the electromagnetic potentials Φ and $\boldsymbol{A}$ from field theory must be replaced by a kind of 'regularized' potentials generated by all the charged particles in the system except the test charge whose motion is being studied. Consequently, the Lorentz force law (2.16) found in textbooks would be replaced by

$$\boldsymbol{F}^{\text{Lor}}_{\text{ext}} = q(\boldsymbol{E}^{\text{ext}} + \boldsymbol{v} \times \boldsymbol{B}^{\text{ext}}), \tag{6.22}$$

because if we use the total potentials Φ and $\boldsymbol{A}$ to obtain the total electromagnetic field, we will end up with a meaningless divergent expression $\boldsymbol{F}^{\mathrm{Lor}} = \infty$, which cannot be used in any equation of motion. The usual literature is both ambiguous and obscure regarding the proper interpretation of the terms in the Lorentz force. A noticeable exception is Wald's advanced text,[10] which states that "*it is understood that **E** and **B** are the external fields.*" I prefer to use the notation $\boldsymbol{E}^{\mathrm{ext}}$ and $\boldsymbol{B}^{\mathrm{ext}}$ here and thereafter to avoid confusion with the true $\boldsymbol{E}$ and $\boldsymbol{B}$ of the field theory.

The main **conclusions** of this chapter are the following:

- Contrary to statements found in mainstream textbooks on classical electrodynamics, the Lorentz force is not "*the force*" that a charge experiences in presence of an electromagnetic field, because there are three different concepts of force in electrodynamics and all of them are physically meaningful.
- The Lorentz force is not the force that we have to use in Hamiltonian or Lagrangian dynamics, except in the particular cases in which $\boldsymbol{F}^{\mathrm{Ham}} = \boldsymbol{F}^{\mathrm{Lor}}$.
- The Lorentz force is not the force associated to the product of mass by acceleration, except when $\boldsymbol{F}^{\mathrm{Lor}} = \boldsymbol{F}^{\mathrm{Newt}}$.
- $\boldsymbol{\pi} = m\gamma\boldsymbol{v}$ is not the momentum of the charged particle.
- The Hamiltonian, Lorentz, and Newtonian forces do not describe the action of a charged particle on itself and the usual Lorentz equation of motion found in textbooks is associated to the external force $\boldsymbol{F}^{\mathrm{Lor}}_{\mathrm{ext}}$ instead of the total force $\boldsymbol{F}^{\mathrm{Lor}}$ predicted by field theory.

—7

RADIATION REACTION AND RENORMALIZATION

In the previous chapter we discussed three definitions of force. However, most textbooks on electrodynamics exclusively use (6.20) for the special case when the speed of the charge is small compared to the speed of light and therefore $\boldsymbol{F}^{\mathrm{Newt}} \approx \boldsymbol{F}^{\mathrm{Lor}}$ (see the mentions on page 72 and the discussion on page 73). For simplicity, we will now focus on this special case and, since there can be no confusion, we will simplify the notation by removing the superscript 'Lor' in this chapter. Furthermore, we also saw in the previous chapter that the true Lorentz force $\boldsymbol{F}$ must be replaced by another force defined in terms of a hypothetical external field, which yields the following equation of motion

$$m\boldsymbol{a} = \boldsymbol{F}^{\mathrm{ext}}. \tag{7.1}$$

This equation of motion with the force law (6.22) is very common in electrodynamics textbooks, but both the equation of motion and the force $\boldsymbol{F}^{\mathrm{ext}} = q(\boldsymbol{E}^{\mathrm{ext}} + \boldsymbol{v} \times \boldsymbol{B}^{\mathrm{ext}})$ are incorrect from the point of view of the theory of field. I will give two reasons why equation (7.1) and the corresponding force are incompatible with a model of electromagnetic phenomena based on the existence of an electromagnetic field.

First of all, there is a single electromagnetic field in this theory. The decomposition of this field into external and self components,

$$\begin{aligned} \boldsymbol{E} &= \boldsymbol{E}^{\text{ext}} + \boldsymbol{E}^{\text{self}} \\ \boldsymbol{B} &= \boldsymbol{B}^{\text{ext}} + \boldsymbol{B}^{\text{self}}, \end{aligned} \tag{7.2}$$

is unjustified, because only $\boldsymbol{E}$ and $\boldsymbol{B}$ have physical meaning in field theory and appear in the Maxwell equations (2.1)–(2.4), and in the Lagrangian (2.17) and the corresponding Hamiltonian (2.18). Consequently, the equation of motion derived from the Lagrangian or the Hamiltonian has to be based on this single electromagnetic field and not in terms of nonexistent "*external fields*"[10] for each charge. The equation of motion (7.1) with the external force law $\boldsymbol{F}^{\text{ext}} = q(\boldsymbol{E}^{\text{ext}} + \boldsymbol{v} \times \boldsymbol{B}^{\text{ext}})$ has no physical justification in the framework of field theory.

We saw in the previous chapter that field theorists identify the quantity $(m\gamma\boldsymbol{v})$ with the momentum of a charged particle in order to associate the Lorentz force with the rate of change of the momentum. We saw then that this identification is not acceptable. We will see now that things are even worse when we consider self-actions. We will use Jackson for this task.[6]

We write again (6.11), but replacing $(m\gamma\boldsymbol{v})$ with $\boldsymbol{\pi}$

$$\frac{\mathrm{d}\boldsymbol{\pi}}{\mathrm{d}t} = \boldsymbol{F}. \tag{7.3}$$

Here the Lorentz force $\boldsymbol{F} = q(\boldsymbol{E} + \boldsymbol{v} \times \boldsymbol{B})$ is given in terms of the electromagnetic field because this equation was derived from the Hamiltonian of the field theory.

According to Jackson, if $\boldsymbol{G}$ is "*the total electromagnetic momentum*" of the field, then the total momentum for the system of a charged particle and the field is $(\boldsymbol{\pi} + \boldsymbol{G})$, and this is a constant quantity

$$\frac{\mathrm{d}\boldsymbol{\pi}}{\mathrm{d}t} + \frac{\mathrm{d}\boldsymbol{G}}{\mathrm{d}t} = 0. \tag{7.4}$$

Jackson considers a single charge for simplicity and identifies $\boldsymbol{\pi}$ as the "*mechanical momentum*", but he immediately states that $\boldsymbol{\pi}$ is not really the momentum of the charge. Jackson first uses the decomposition (7.2) in the right hand side of the equation of motion (7.3), obtaining $\boldsymbol{F} = \boldsymbol{F}^{\mathrm{ext}} + \boldsymbol{F}^{\mathrm{self}}$, next moves the self-force to the left-hand side of the equation of motion, and finally reinterprets this side as the rate of change of the momentum associated with the charged particle. That is,

$$\underbrace{\left(\frac{\mathrm{d}\boldsymbol{\pi}}{\mathrm{d}t} - \boldsymbol{F}^{\mathrm{self}}\right)}_{\text{CHARGE}} = \boldsymbol{F}^{\mathrm{ext}}. \tag{7.5}$$

Therefore, according to Jackson, "*the particle's momentum is partly mechanical, but with an electromagnetic contribution*". This electromagnetic contribution is given by minus the time-integral of the Lorentzian self-force.

We saw in the previous chapter that $\boldsymbol{\pi} = (m\gamma\boldsymbol{v})$ cannot be interpreted as the momentum of a charged particle. Our argument was based on the consistency of the phase space structure. Now we see that, even ignoring these consistency requirements, $\boldsymbol{\pi}$ still cannot be the momentum of the particle once we consider self-forces. But then, if $\boldsymbol{\pi}$ no longer represents the momentum of the charged particles, the rate of change (7.4) for the total momentum of a system of charges and field must be reinterpreted as

$$\underbrace{\left(\frac{\mathrm{d}\boldsymbol{\pi}}{\mathrm{d}t} - \boldsymbol{F}^{\mathrm{self}}\right)}_{\text{CHARGE}} + \underbrace{\left(\frac{\mathrm{d}\boldsymbol{G}}{\mathrm{d}t} + \boldsymbol{F}^{\mathrm{self}}\right)}_{\text{FIELD}} = 0, \tag{7.6}$$

in which case $\boldsymbol{G}$ is not the electromagnetic momentum of the field, but Jackson does not come to this logical conclusion from his own line of development and simply changes the subject at this point. We have analyzed Jackson's textbook, but similar inconsistencies and mistakes are found in the rest of the literature.

We might imagine from the above discussion that (7.5) is the proper equation of motion for a charged particle and that this equation reduces to (7.1) for very low velocities, but both equations are still incorrect. We will continue the discussion with (7.1) for simplicity.

We have just seen the first reason why (7.1) is incompatible with field theory: the electromagnetic field is given by $\boldsymbol{E}$ and $\boldsymbol{B}$ and not by $\boldsymbol{E}^{\text{ext}}$ and $\boldsymbol{B}^{\text{ext}}$. Another reason is that, according to the laws of classical field theory, accelerated charges radiate. This radiation takes away energy from the charge, which must be done at the expense of reducing the kinetic energy of the charge. Under the influence of any external force, a charged particle accelerates less than a neutral particle of the same mass, and physicists[5,6,9] simply state that the equation of motion (7.1) needs to be modified. Quoting Jackson:[6]

> *"To account for this radiative energy loss and its effect on the motion of the particle we modify the Newton equation 6.5 by adding a radiative reaction force* $\boldsymbol{F}^{\text{rad}}$*:*
> $$m\boldsymbol{a} = \boldsymbol{F}^{\text{ext}} + \boldsymbol{F}^{\text{rad}}."$$

At this point, $\boldsymbol{F}^{\text{rad}}$ is derived from the Larmor formula for the rate of energy loss when the speed of the charge is very small compared to the speed of light. The Larmor formula for this particular case is

$$\frac{\mathrm{d}E}{\mathrm{d}t} = \frac{q^2}{6\pi\epsilon_0 c^3}\,\boldsymbol{a}\cdot\boldsymbol{a}. \tag{7.7}$$

We determine the form of $\boldsymbol{F}^{\text{rad}}$ by requiring that the average work done by this force on the charged particle be equal to the negative of the energy radiated on average

$$\int \boldsymbol{F}^{\text{rad}} \cdot \boldsymbol{v}\,\mathrm{d}t = -\int \frac{q^2}{6\pi\epsilon_0 c^3}\,\boldsymbol{a}\cdot\boldsymbol{a}\,\mathrm{d}t. \tag{7.8}$$

Assuming that the motion is periodic, the second integral can be integrated by parts to obtain the following force

$$\boldsymbol{F}^{\text{rad}} = \frac{q^2}{6\pi\epsilon_0 c^3}\frac{\mathrm{d}\boldsymbol{a}}{\mathrm{d}t}. \tag{7.9}$$

Note: The term radiation-reaction is used to describe the effect of $\boldsymbol{F}^{\text{rad}}$ on a charge, but this term is a misnomer. We should really call this effect "*the field reaction*", as Griffiths points out.[5] We will continue to use the term radiation-reaction because this is the standard terminology in the literature.

The resulting equation of motion

$$m\boldsymbol{a} = \boldsymbol{F}^{\text{ext}} + \frac{q^2}{6\pi\epsilon_0 c^3}\frac{\mathrm{d}\boldsymbol{a}}{\mathrm{d}t} \tag{7.10}$$

is the famous Abraham-Lorentz equation, named after Abraham modified the Lorentz equation. However, this equation of motion is not correct and raises many problems. One of the difficulties with this equation is that the radiation-reaction force depends on the derivative of the acceleration. Dividing both sides of the equation by m and introducing the shorthand $\tau = q^2/(6\pi\epsilon_0 mc^3)$ to simplify the notation, yields

$$\boldsymbol{a} = \frac{\boldsymbol{F}^{\text{ext}}}{m} + \tau\frac{\mathrm{d}\boldsymbol{a}}{\mathrm{d}t}. \tag{7.11}$$

The general solution of this equation of motion is

$$\boldsymbol{a}(t) = \exp\left(\frac{t}{\tau}\right)\boldsymbol{a}(0) - \frac{1}{m\tau}\int_0^t \exp\left(\frac{t-s}{\tau}\right)\boldsymbol{F}^{\text{ext}}(s)\,\mathrm{d}s, \tag{7.12}$$

with $\boldsymbol{a}(0)$ the initial acceleration of the charged particle and s an integration variable. Suppose there are no external forces, then $\boldsymbol{F}^{\text{ext}} = 0$ and the solution reduces to

$$\boldsymbol{a}(t) = \exp\left(\frac{t}{\tau}\right)\boldsymbol{a}(0), \tag{7.13}$$

which implies that the acceleration $\boldsymbol{a}(t)$ increases indefinitely unless the initial acceleration $\boldsymbol{a}(0)$ is zero. This is the "*runaway solution*" to the Abraham-Lorentz equation and is completely unacceptable. How could a particle not subjected to any external force accelerate continuously?

If we set $\boldsymbol{a}(0) = \boldsymbol{0}$ as an initial-boundary condition we eliminate the runaways, but then the equation of motion is only valid for particles with constant initial velocity; that is, it is only valid for particles to which no external forces are initially applied.

Runaway solutions can also be avoided by imposing the final-boundary condition that the acceleration vanishes at very long times; that is, $\boldsymbol{a}(\infty) = \boldsymbol{0}$. The solution to the Abraham-Lorentz equation is now

$$\boldsymbol{a}(t) = \frac{1}{m\tau}\int_t^\infty \exp\left(\frac{t-s}{\tau}\right)\boldsymbol{F}^{\text{ext}}(s)\text{d}s, \tag{7.14}$$

but it could be said that the cure is worse than the disease, because the acceleration depends on values of the external force $\boldsymbol{F}^{\text{ext}}(s)$ at future times $s > t$. If we do apply an external force, the charged particle starts to respond before the force acts on it. This is what physicists call the "*preacceleration solution*".

Runaways violate conservation of energy and preacceleration violates causality, and "*both offend common sense*",[33] which has forced physicists to search for an alternative equation of motion.

Note: Preacceleration violates causality only in the context of a field-model of interactions based on retarded potentials with signals traveling backward in time, which is a misconception as we saw in chapter 5. A real difficulty with the preacceleration solution is that it requires all accelerations to cease in the distant future. There is no physical reason why all charged particles should tend asymptotically to an inertial regime in which all velocities are constant.

To obtain an alternative equation, let us return to the Abraham-Lorentz equation with the simplified notation; that is, equation (7.11). If we differentiate both sides

$$\frac{\mathrm{d}\boldsymbol{a}}{\mathrm{d}t} = \frac{1}{m}\frac{\mathrm{d}\boldsymbol{F}^{\mathrm{ext}}}{\mathrm{d}t} + \tau\frac{\mathrm{d}^2\boldsymbol{a}}{\mathrm{d}t^2}, \tag{7.15}$$

replace this result back into the right-hand side of (7.11) and finally ignore the term proportional to τ^2, we obtain the following equation of motion

$$\boldsymbol{a} = \frac{\boldsymbol{F}^{\mathrm{ext}}}{m} + \frac{\tau}{m}\frac{\mathrm{d}\boldsymbol{F}^{\mathrm{ext}}}{\mathrm{d}t}. \tag{7.16}$$

This is the Landau-Lifshitz equation, an equation of motion without runaway solutions or acausal behavior. It is sometimes considered *"a sensible alternative to the Abraham-Lorentz equation for the classical regime of small radiative effects"*,[6] but why would we trust an approximation like (7.16) that ignores a term proportional to τ^2, and not the equation (7.11) that it approximates?

Poisson argues that *"a point particle cannot be taken too literally in a classical context; it must always be considered as an approximation to a nonsingular, and extended, charge distribution"*.[34] Consequently, he starts with an extended charge distribution and claims that the point-particle description is only valid for distances much larger than the characteristic size of the charge distribution. This is equivalent to stating that the Abraham-Lorentz equation is only valid for times much longer than τ.

Poisson then uses the same replacement procedure we used to derive the Landau-Lifshitz equation and claims that it has the same degree of accuracy as the original Abraham-Lorentz equation in the sense that they are both accurate up to terms of order τ^2. However, with his claim that the Landau-Lifshitz equation "*is much better suited to govern the motion of a point charge*" he is implicitly admitting that the Abraham-Lorentz equation, which is a direct consequence of field theory, is flawed; although Poisson does not admit it explicitly.

Furthermore, recent studies have shown that the Landau-Lifshitz equation has its own set of difficulties:[33] "*In some cases the Landau-Lifshitz formula is a good approximation. But it seems to us an overstatement to represent the Landau-Lifshitz equation as 'physically correct'*". Even if those difficulties could someday be overcome, I fully agree with Griffiths, Proctor, and Schroeter that replacing (7.11) with (7.16) raises "*the delicate question of how an approximation can be considered more accurate than the original*".[33]

For all of the above reasons, some physicists directly replace the Abraham-Lorentz equation (7.10) with a difference-delay equation where the charges are modeled as uniformly charged spherical shells of radius λ

$$m\boldsymbol{a} = \boldsymbol{F}^{\text{ext}} + \frac{q^2}{12\pi\epsilon_0\lambda^2 c}\left[\boldsymbol{v}\left(t - \frac{2\lambda}{c}\right) - \boldsymbol{v}(t)\right]. \qquad (7.17)$$

This equation is expected to be free of the pathologies that affect other equations if we assume a radius λ much larger than the Lorentz radius λ_e, whose value is $2.817940 \cdot 10^{-15}$m. But be careful, because assuming this value for λ is equivalent to considering an electron to be about 10,000 times bigger than the upper bound radius inferred from measurements of its anomalous magnetic moment.

Note: The Lorentz radius λ_e or Thomson scattering length is often known as the classical electron radius, and is interpreted as the radius such that all the physical mass of the electron has an electromagnetic origin. The term classical electron radius is a misnomer, because electrons are treated as pointlike objects both classically and quantumly.

Even if we assume unphysical sizes for the electrons, equation (7.17) is not useful to describe motion, because the acceleration at any instant t depends on the value of the velocity of the particle at early times $(t - 2\lambda/c)$, which makes it impossible to define a dynamics in terms of a finite number of variables. In reality, this problem is already hidden in the expression for the Lorentz force, when its value is calculated by means of the retarded potentials of the field theory, but we will discuss the impossibility of defining a dynamics in the next chapter. Moreover, the mathematical proof of the nonexistence of runaways on (7.17) is disputed.[35] After all the efforts to solve the problems with the Abraham-Lorentz equation, we are back to the beginning.

We obtained the Abraham-Lorentz equation (7.10) by adding a radiative-reaction force to (7.1). While plausible, our derivation is certainly neither rigorous nor fundamental. We will now focus on how to systematically obtain, and from first principles, the reaction of the charged particle to its own radiation field.

We consider again equation (7.1), but replacing $\boldsymbol{F}^{\mathrm{ext}}$ with the total Lorentz force $\boldsymbol{F}$

$$m\boldsymbol{a} = \boldsymbol{F}. \tag{7.18}$$

As we have seen in this and the previous chapter, the law of conservation of energy is violated when we use only the external Lorentz force in the equation of motion, while the motion is ill-defined (divergences) when we use the total Lorentz force. A more adequate equation should be somewhere in between those two extremes.

We begin by applying the decomposition (7.2) to (7.18),

$$m\boldsymbol{a} = \boldsymbol{F}^{\text{ext}} + \boldsymbol{F}^{\text{self}}. \tag{7.19}$$

The net Lorentz force acting on the charge is the sum of the external force given by (6.22) and the force exerted by the particle on itself $\boldsymbol{F}^{\text{self}} = q(\boldsymbol{E}^{\text{self}} + \boldsymbol{v} \times \boldsymbol{B}^{\text{self}})$. The usual method of calculating the self-force and then deriving the radiation reaction force from it is to use the retarded potentials for the self-force, but first decomposing those potentials into a singular part (S) and a radiative remainder (R), also called regular,

$$\begin{aligned} \Phi^{\text{ret}} &= \Phi^{(S)} + \Phi^{(R)} \\ \boldsymbol{A}^{\text{ret}} &= \boldsymbol{A}^{(S)} + \boldsymbol{A}^{(R)}, \end{aligned} \tag{7.20}$$

and then *postulating* that the radiative potential "*and it alone, exerts a force on the particle*".[36] In a curved spacetime context (that is, when we combine electrodynamics with general relativity), this assumption is called the Detweiler-Whiting axiom: "*the singular field exerts no force on the particle (it merely contributes to the particle's inertia); the entire self-force arises from the action of the radiative field*".[36]

Not only are *additional postulates* required to make sense of the equation of motion of a charged particle in field theory, but the radiative fields used to calculate the radiation-reaction force depend on the advanced potentials.

The advanced potentials are usually rejected on grounds of causality violation, as we saw in chapter 5. Under what criteria can we reject the advanced potentials for the external component of the forces but admit them for the self-forces? The situation is even more awkward in a curved spacetime context, because the delayed potentials are not evaluated at a single retarded or advanced point. The retarded potential of a charged particle in curved spacetime depends on the history of the particle before the retarded time, and the corresponding advanced potential depends on the history after the advanced time. Moreover, the radiative potentials also originate outside the light cones, as shown in figure 7.1.

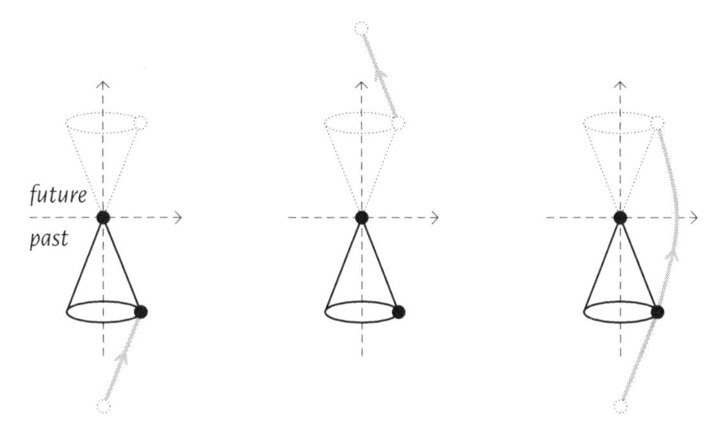

Figure 7.1: Retarded potentials (left), advanced potentials (center), and radiative potentials (right) in curved spacetime.

The usual method of calculating the self-force and then deriving from it the radiation reaction term involve rather cumbersome calculations and are conceptually obscure because they invoke additional postulates and are based on hidden assumptions. We will offer a simpler and more direct approach.

We begin by splitting the self-force $\boldsymbol{F}^{\text{self}}$ in (7.19) into two halves and then we add (+) and subtract (−) to the equation of motion one-half of an alternative self-force $\boldsymbol{F}^{\text{self}}_{\text{adv}} = q(\boldsymbol{E}^{\text{self}}_{\text{adv}} + \boldsymbol{v} \times \boldsymbol{B}^{\text{self}}_{\text{adv}})$ defined in terms of the *advanced potentials*. The result is

$$m\boldsymbol{a} = \boldsymbol{F}^{\text{ext}} + \frac{1}{2}\left(\boldsymbol{F}^{\text{self}} - \boldsymbol{F}^{\text{self}}_{\text{adv}}\right) + \frac{1}{2}\left(\boldsymbol{F}^{\text{self}} + \boldsymbol{F}^{\text{self}}_{\text{adv}}\right). \tag{7.21}$$

This equation is exactly equivalent to (7.19) because we have only added a null term $(1/2)(\boldsymbol{F}^{\text{self}}_{\text{adv}} - \boldsymbol{F}^{\text{self}}_{\text{adv}})$ to it. This equation is physically as meaningless as (7.18) and (7.19) due to the presence of divergences; the equation is $m\boldsymbol{a} = \infty$. However, we have now been able to confine these divergences within the $(\boldsymbol{F}^{\text{self}} + \boldsymbol{F}^{\text{self}}_{\text{adv}})$ term.

Keeping the same low-velocity approximation as in the rest of this chapter for convenience, we can show that the difference between retarded and advanced self-forces gives the Abraham radiation-reaction force; that is,

$$\frac{1}{2}\left(\boldsymbol{F}^{\text{self}} - \boldsymbol{F}^{\text{self}}_{\text{adv}}\right) = \frac{q^2}{6\pi\epsilon_0 c^3}\frac{\mathrm{d}\boldsymbol{a}}{\mathrm{d}t}. \tag{7.22}$$

This term is finite because the minus sign cancels the divergences from the two self-forces and always gives a radiation-reaction force. We can call it the radiation-reaction term. If we had not applied the low-velocity approximation, we would have obtained a generalization of the Abraham force.

Note: Being rigorous, $(\infty - \infty)$ is undefined in mathematics. What we do is to consider a charge with finite radius λ, perform all the algebraic manipulations with finite quantities and, at the end of the mathematical procedure, take the limit when the radius reaches zero.

Although the radiation-reaction term is a finite quantity, the sum of the retarded and the advanced self-forces in (7.21) is divergent and proportional to the acceleration

$$\frac{1}{2}\left(\boldsymbol{F}^{\text{self}} + \boldsymbol{F}^{\text{self}}_{\text{adv}}\right) = -\lim_{\lambda\to 0}\frac{q^2}{6\pi\epsilon_0 c^2\lambda}\boldsymbol{a}. \tag{7.23}$$

We will call this the mass-renormalization term; the reason will become evident in a moment. Moving this term to the left-hand side of (7.21), we obtain

$$\left(m + \lim_{\lambda\to 0}\frac{q^2}{6\pi\epsilon_0 c^2\lambda}\right)\boldsymbol{a} = \boldsymbol{F}^{\text{ext}} + \frac{q^2}{6\pi\epsilon_0 c^3}\frac{\mathrm{d}\boldsymbol{a}}{\mathrm{d}t}. \tag{7.24}$$

This looks like the Abraham-Lorentz equation (7.10), but with a modified mass on the left side. Recall that equation (7.24) has been derived from the first principles of the field theory of electrodynamics, while (7.10) was only derived heuristically by adding an *ad hoc* radiation-reaction force to an external-field Newton-

Lorentz equation (7.1) that is incorrect from the point of view of field theory. We initially followed the heuristic method because this is how the Abraham-Lorentz equation is presented in most textbooks.

Theoretical physicists affirm that (7.24) is the correct equation of motion for low velocities and that the real mass of the charged particle is not m but the modified mass

$$M = m + \delta m = m + \lim_{\lambda \to 0} \frac{q^2}{6\pi\epsilon_0 c^2 \lambda}. \tag{7.25}$$

That is, they affirm that both the mass m and the correction δm are *unobservable* quantities. The quantity M is what we measure in the laboratory. This interpretation produces two serious theoretical difficulties.

The first difficulty with (7.25) is that the correction to the original mass m diverges because $\delta m \to \infty$ when $\lambda \to 0$. In order to keep a finite M compatible with experimental findings for point charges such as the electron, the original mass m must be also divergent. Moreover, $(\infty + \infty) = \infty$, which implies that the original mass *m must be negative!* Effectively, the equation of motion (7.24) only makes sense if

$$\begin{aligned} m &= -\infty \\ \delta m &= +\infty \\ M = m + \delta m &= \text{ measured mass.} \end{aligned} \tag{7.26}$$

One might imagine that these difficulties are the result of pushing classical electrodynamics too far into the range of the very small, and that the difficulties would disappear with a proper quantum electrodynamics treatment. However, as Griffiths admits, "*this awkward problem persists in quantum electrodynamics, where it is 'swept under the rug' in a process known as mass renormalization*".[5]

The process of isolating all the divergences resulting from the electromagnetic interaction of a charge with itself and then absorbing these divergences into a redefinition of the mass of the particle, replacing m with M, is known as mass renormalization. This is why we referred to (7.23) as a mass-renormalization term.

The second difficulty with (7.25) is that m appears in both the Lagrangian (2.17) and the Hamiltonian (2.18) of the field theory. The actual mass M of the charged particles only appears on the equations of motion after we renormalize these equations to remove the unphysical divergences. How is it possible that the real masses of the particles do not appear in either the Lagrangian or the Hamiltonian of field theory? The answer is obvious: the field theory of electromagnetism is a flawed theory.

Note: Contrary to what is often believed, the only problem with renormalization is not the presence of divergences in the fundamental equations. Even if this nonphysical mass m was finite, it would be necessary to apply a renormalization procedure to replace it with the actual mass M compatible with laboratory measurements. Field theory is defined in terms of unphysical parameters q and m that are not the actual charges and masses of the particles.

It is worth making a small parenthesis now. We constantly make a distinction between physical and nonphysical masses, but many physicists refer to them as "*experimental*" and "*theoretical*" masses, respectively. The terminology used by them is not trivial, but it reflects a current attitude in theoretical physics that began in the 1970s and that attempts to minimize fundamental questions. The Nobel prize laureate Salam commented in 1972:

> "*Field-theoretic infinities —first encountered in Lorentz's computation of electron self-mass— have persisted in classical electrodynamics for seventy [years] and in quantum electrodynamics for some thirty-five years. These long years of frustration have left in the subject a curious affection for the infinities and a passionate be-*

> *lief that they are an inevitable part of nature; so much so that even the suggestion of a hope that they may, after all, be circumvented —and finite values for the renormalization constants computed— is considered 'irrational'."*

Infinities and negative masses are an inevitable part of the use of fields to describe interactions between particles, but they are not an inevitable part of Nature. Field theorists are confusing the model with reality. There are no infinities or negative masses in the action-at-a-distance model of interactions. Thus, we will continue to refer to M as the actual mass and to m as an nonphysical mass.

In the first part of this chapter, we wrote equations like (7.10), using the symbol m to denote the mass of the charged particle. We did this because we were just reproducing results found in electrodynamics textbooks where the technical subtleties of mass-renormalization are ignored. If we insist on using m to denote the real mass in the equations of motion, then we would rewrite both the Lagrangian (2.17) and the Hamiltonian (2.17), and use another symbol for the nonphysical masses that appear in them. If we want to keep m in the Lagrangian and the Hamiltonian, then we would use another symbol for the actual masses that appear in the equations of motion; we have used M in the second part of this chapter. Be warned that standard textbooks will most likely use the same symbol m to denote both real and nonphysical masses, since textbooks ignore mass renormalization.

In this chapter, we have considered only the special case in which the speed of the charge is small compared to the speed of light, but the difficulties with the equations of motion persist in the high-velocity regime. In this regime, the equation (7.24) is replaced by the Abraham-Lorentz-Dirac equation, which we will not write here.

In relation to the difficulties with the Abraham-Lorentz-Dirac equation, Griffiths asserts[5] that the difficulties perhaps "*are telling us that there can be no such thing as a point charge in classical electrodynamics*", Poisson is more categorical and affirms that these "*difficulties stem from the basic observation that point particles cannot be given a fully consistent treatment in a classical theory of electromagnetism*".[34]

This is all incorrect. Coulomb theory is a classical theory fully capable of describing point particles, and the same is true of the special-relativistic theories of Fokker, Wheeler, and Feynman. The origin of the difficulties is not in a classical description, but rather the difficulties are a consequence of the use of fields to describe interactions. As we will see in appendix A, a proper action-at-a-distance model of electromagnetic interactions is equivalent to a field model plus renormalization corrections.

Recall that some physicists promote the difference-delay equation (7.17), where charges are modeled as uniformly charged spherical shells of finite radius, as an equation of motion with no pathological solutions and that leads to a completely acceptable formulation of classical electrodynamics. However, as we already mentioned on page 83, such claims have not yet been proven. Furthermore, the mathematical techniques used to analyze the behavior and properties of (7.17) are not available for its generalization to arbitrary physical regimes.[35]

The main **conclusions** of this chapter are the following:

- The usual Lorentzian equation of motion of the field theory is incomplete, and the alternative equations derived by Abraham and by Landau and Lifshitz, as well as the equation of motion for a charged spherical shell, have serious difficulties, including the violation of the principles of causality and conservation of energy.

- The alternative equations cannot be directly derived from the standard Lagrangian or Hamiltonian of the field theory of electromagnetism, unless additional postulates are invoked.
- The mass m that appears in the field theory is a divergent negative quantity, which must be replaced by the real experimental mass by means of a renormalization procedure.
- We learned that the term "*radiation-reaction*" is a misnomer, we also learned that the advanced potentials are usually rejected for violation of causality, but then used to obtain radiation forces.
- Contrary to myth, field theory cannot completely describe the motion of a point particle, but the problem is not in the concept of point particles, but in the use of fields to describe interactions.

—8

MANY-BODY MOTION

After verifying in the previous chapter that the field theory of electromagnetism cannot produce a complete and satisfactory equation of motion for a single charged particle like the electron, we can infer that this theory cannot describe the motion of a system of many charges. Indeed, if a complete and satisfactory dynamics does not exist for a single body, how could it exist for two or more bodies? This is pure logic, but the problems with field theory do not end with radiation-reaction. As we will show in this chapter, field theory cannot describe the motion of a system of two or more interacting particles, even if we ignore self-actions.

A cursory review of the main treatises on classical electrodynamics[5,6,8] and on classical mechanics[37–41] reveals that some authors only give Lagrangians and Hamiltonians for a single "*charged particle in an external electromagnetic field*", while others consider the case of many particles but only provide an approximate description; for example, one valid exclusively for very small speeds.[9]

A notable exception to this rule is the textbook by Landau and Lifshitz,[7] which gives the Lagrangian (2.17) for a system of charges plus the electromagnetic field. Nevertheless, despite appearances, (2.17) and the corresponding Hamiltonian (2.18) are not valid for the description of many-body motion.

Consider the simplest case of two bodies, if we set $N = 2$ in (2.18) and solve the Hamilton equations, we obtain a pair of equations of motion, one for each charged particle. To continue the discussion, we will convert those equations to the more common Lorentzian form (see chapter 6 for details). The result is

$$\frac{\mathrm{d}(m_1\gamma_1\boldsymbol{v}_1)}{\mathrm{d}t} = q_1(\boldsymbol{E} + \boldsymbol{v}_1 \times \boldsymbol{B}) \tag{8.1}$$

and

$$\frac{\mathrm{d}(m_2\gamma_2\boldsymbol{v}_2)}{\mathrm{d}t} = q_2(\boldsymbol{E} + \boldsymbol{v}_2 \times \boldsymbol{B}). \tag{8.2}$$

As we saw in the previous chapter, the first difficulty is the divergences introduced by the self-action of the electromagnetic field. Replacing $\boldsymbol{E}$ by $\boldsymbol{E}^{\text{ext}}$ and $\boldsymbol{B}$ by $\boldsymbol{B}^{\text{ext}}$ is the standard procedure used in the literature for the study of a single particle, when one wants to ignore self-actions effects. For the two-body case, we can remove the divergences by the *ad hoc* procedure of assuming that the field acting on the first particle is generated only by the second particle and vice versa.

We already know the difficulties that affect to the equation of motion when we replace the field by an external field, but here I want to point out another difficulty when we are studying more than one particle: the above replacement implies that we have gone from a single electromagnetic field filling the whole space to N fields filling only certain regions. In our working example, $N = 2$ and we will have to deal with two electromagnetic fields.

Thus, the $\boldsymbol{E}^{\text{ext}}$ and $\boldsymbol{B}^{\text{ext}}$ for the equation of motion of the first particle are not the same $\boldsymbol{E}^{\text{ext}}$ and $\boldsymbol{B}^{\text{ext}}$ that we will have to utilize for the second particle. Using subscripts to differentiate each field, the new equations of motion are

$$\frac{\mathrm{d}(m_1\gamma_1\boldsymbol{v}_1)}{\mathrm{d}t} = q_1(\boldsymbol{E}_1^{\text{ext}} + \boldsymbol{v}_1 \times \boldsymbol{B}_1^{\text{ext}}) \tag{8.3}$$

and

$$\frac{\mathrm{d}(m_2\gamma_2\boldsymbol{v}_2)}{\mathrm{d}t} = q_2(\boldsymbol{E}_2^{\text{ext}} + \boldsymbol{v}_2 \times \boldsymbol{B}_2^{\text{ext}}). \tag{8.4}$$

Obviously, this split into two external fields goes against the main assumption of the field theory that there is a single electromagnetic field common for all the interacting particles. We will ignore this difficulty and we will also ignore the additional problems related to violating energy conservation. What I want to emphasize in this chapter is that, even if we ignore all this, equations (8.3) and (8.4) do not give a physical theory for the two-body system of charges. Now we will see why.

Consider the retarded potentials (5.1) and (5.2), the external electromagnetic field acting at instant t on one of the charged particles, say the first particle, depends on the position and velocity of the second particle at the earlier time $t^{\text{ret}} = (t - |\boldsymbol{r}_1 - \boldsymbol{r}_2'|/c)$. Likewise, the motion of the second particle at the earlier time t^{ret} is due entirely to the position and velocity of the first particle at a much earlier time $t^{\text{retret}} = (t^{\text{ret}} - |\boldsymbol{r}_2' - \boldsymbol{r}_1|/c)$.

In other words, we need to know the position $\boldsymbol{r}_2(t^{\text{ret}})$ and the velocity $\boldsymbol{v}_2(t^{\text{ret}})$ for the second charge if we want to calculate the motion of the first charge at instant t, but in turn we need to know the position $\boldsymbol{r}_1(t^{\text{retret}})$ and the velocity $\boldsymbol{v}_1(t^{\text{retret}})$ of the first charge in order to obtain $\boldsymbol{r}_2(t^{\text{ret}})$ and $\boldsymbol{v}_2(t^{\text{ret}})$ for the second charge, and so on. Figure 8.1 illustrates this.

This dependence of the motion of one of the charged particles on the earlier motion of the other particle generates an *endless causal chain* between the two particles, which we represent as

$$\underbrace{\cdots \to \boldsymbol{r}_1(t^{\text{retret}}), \boldsymbol{v}_1(t^{\text{retret}}) \to \boldsymbol{r}_2(t^{\text{ret}}), \boldsymbol{v}_2(t^{\text{ret}})}_{\text{PAST}} \to \underbrace{\boldsymbol{r}_1(t), \boldsymbol{v}_1(t)}_{\text{PRESENT}} \tag{8.5}$$

and

$$\underbrace{\cdots \to \boldsymbol{r}_2(t^{\text{retret}}), \boldsymbol{v}_2(t^{\text{retret}}) \to \boldsymbol{r}_1(t^{\text{ret}}), \boldsymbol{v}_1(t^{\text{ret}})}_{\text{PAST}} \to \underbrace{\boldsymbol{r}_2(t), \boldsymbol{v}_2(t)}_{\text{PRESENT}} . \tag{8.6}$$

This endless causal chain implies an *infinite number of dynamical degrees of freedom*; that is, an infinite collection of positions and velocities.

Note: One might believe that the causal chain cannot extend into the past forever because the big bang model used in cosmology supposedly introduced a beginning for time, but this is a misconception propagated by physicists like Hawking. First of all, this model is not valid for the first stages of cosmological evolution since it does not consider quantum gravity effects, much less describes the hypothetical initial singularity. Therefore, Hawking's claim that time started with the big bang is unfounded. Second, there are recent cosmological models without singularity, which have been developed to solve the homogeneity, isotropy, flatness, and monopole problems of the big bang inflationary cosmology. In such models, the universe is eternal and evolves cyclically through an eternal series of cosmological expansions and contractions.[42] I do not support any of the cosmological models mentioned here, I simply wanted to point out a fairly common misconception.

The requirement of knowing an infinite number of positions and velocities prohibits, for all practical purposes, the formulation of the two-body dynamics as an *initial value problem* in which the initial conditions are the instantaneous positions and velocities of all the charged particles involved (what mathematicians call the Cauchy data). Instead, we obtain a theory in which the initial conditions comprise an infinite regress of segments of particle trajectories.

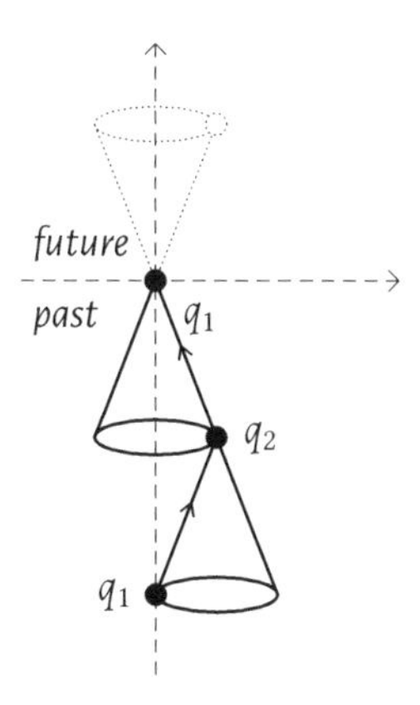

Figure 8.1: Causal chain at retarded times for two interacting charges. Only two iterations are shown –see (8.5)–, but field theory requires an infinite chain.

The aforementioned requirement is often taken as a 'proof' that the electromagnetic field is necessary to describe the motion of a system of interacting charges, since a field has an *"infinite number of degrees of freedom"*.[6] However, adding a field as intermediary of the interactions does not improve the situation, because the number of variables required to describe the system of charges and field remains infinite. From a practical point of view, there is no real difference between the situation of having to measure an infinite number of positions and velocities and the alternative situation of having to measure a finite number of positions and velocities for the charges plus an infinite number of field coordinates.

Before continuing our discussion of many-body motion, I want to make a brief ontological observation here. It is often claimed that fields have been observed. This is not true. *Since fields are infinite systems, we will never be able to measure or detect them*. The best we can do is observe part of a field. However, no one has measured a field locally. When a physicist or engineer claims to have measured an electric field at a given point in space, what he has really done is put a test charge there, measure its acceleration, and then infer the existence of a field. Similar objections apply to statements about "*measurements of the Earth's magnetic field*"[6] and about "*how to measure magnetic fields*".[8] Nobody measures the electromagnetic field as such.

Most standard textbooks ignore many-body motion, but Trump and Schieve address the inability of field theory to fully describe it:[43]

> "*As mentioned above, relativistic field theory is essentially a one-body theory, in that it is incapable of describing the N-body system of point particles, except in certain limits.*"

Note: I do not completely agree with their statement, because the so-called relativistic field theory is also incapable of describing the motion of a single body once we include radiation-reaction effects, as we showed in the previous chapter.

But why cannot field theory completely describe the motion of many bodies? Jackson considers the retardation of the interactions to be the reason why classical electrodynamics cannot do it:[6]

> "*If we now consider the problem of a conventional Lagrangian description of the interaction of two or more charged particles with each other, we see that it is possible only at nonrelativistic velocities. The Lagrangian is supposed to be a function of the instantaneous velocities and coordinates of all the particles. When the finite velocity of propagation of electromagnetic fields is taken into account, this is no longer possible, since the values of the potentials at one particle due to the other particles depend on their state of motion at 'retarded' times. Only when the retardation effects can be neglected is a Lagrangian description in terms of instantaneous positions and velocities possible.*"

However, the existence of the Darwin Lagrangian shows that his argument is not entirely right. Although it is often stated categorically in the literature that the Darwin Lagrangian neglects retardation effects,[43] we will show that retardation is included up to $(1/c)^2$ order.

The complete Darwin Lagrangian for N charges is

$$L^{\text{Darw}} = \sum_i^N \frac{m_i \boldsymbol{v}_i^2}{2} + \frac{m_i \boldsymbol{v}_i^4}{8c^2} - \kappa \sum_i^N \sum_{j \neq i}^N \frac{q_i q_j}{2r_{ij}} \left[1 - \frac{\boldsymbol{v}_i \boldsymbol{v}_j}{2c^2} - \frac{(\boldsymbol{v}_i \boldsymbol{r}_{ij})(\boldsymbol{v}_j \boldsymbol{r}_{ij})}{2c^2 r_{ij}^2} \right], \tag{8.7}$$

with $\kappa = 1/(4\pi\epsilon_0)$ again the Coulomb constant. This Lagrangian depends on the positions and velocities of the charges at the present time, which has led many physicists to believe that Darwin ignored retardation effects in his derivation.

The Darwin Lagrangian consists of kinetic and interaction terms, but we are only interested in the interaction term

$$L_{\text{inter}}^{\text{Darw}} = -\kappa \sum_i^N \sum_{j \neq i}^N \frac{q_i q_j}{2r_{ij}} \left[1 - \frac{\boldsymbol{v}_i \boldsymbol{v}_j}{2c^2} - \frac{(\boldsymbol{v}_i \boldsymbol{r}_{ij})(\boldsymbol{v}_j \boldsymbol{r}_{ij})}{2c^2 r_{ij}^2} \right]. \tag{8.8}$$

If we use the mathematical identity

$$\left[1 - \frac{\boldsymbol{v}_i \boldsymbol{v}_j}{2c^2} - \frac{(\boldsymbol{v}_i \boldsymbol{r}_{ij})(\boldsymbol{v}_j \boldsymbol{r}_{ij})}{2c^2 r_{ij}^2} \right] = \left(1 - \frac{\boldsymbol{v}_i \boldsymbol{v}_j}{c^2} \right) + \left[\frac{\boldsymbol{v}_i \boldsymbol{v}_j}{2c^2} - \frac{(\boldsymbol{v}_i \boldsymbol{r}_{ij})(\boldsymbol{v}_j \boldsymbol{r}_{ij})}{2c^2 r_{ij}^2} \right], \tag{8.9}$$

then we can interpret the Darwin interaction term as the combination of an instantaneous term with a retardation correction term of order $(1/c)^2$

$$L_{\text{inter}}^{\text{Darw}} = -\kappa \sum_i^N \sum_{j \neq i}^N \underbrace{\frac{q_i q_j}{2r_{ij}} \left(1 - \frac{\boldsymbol{v}_i \boldsymbol{v}_j}{c^2} \right)}_{\text{INSTANTANEOUS}} + \underbrace{\frac{q_i q_j}{2r_{ij}} \left[\frac{\boldsymbol{v}_i \boldsymbol{v}_j}{2c^2} + \frac{(\boldsymbol{v}_i \boldsymbol{r}_{ij})(\boldsymbol{v}_j \boldsymbol{r}_{ij})}{2c^2 r_{ij}^2} \right]}_{\text{RETARDATION}}. \tag{8.10}$$

We will show in a moment why the last term describes retardation, but first of all I want to emphasize that the Lagrangian (8.7) cannot be obtained from the Lagrangian (2.17) for a system of charges and electromagnetic field unless we remove the divergences by *adding renormalization counterterms* to (2.17), perform additional algebraic operations to express the modified Lagrangian in the appropriate dynamical form, introduce suitable boundary conditions for the field, and finally reinterpret some interaction terms in the resulting Lagrangian.

For the purposes of this chapter, we will follow the *heuristic derivation* provided by Landau and Lifshitz.[7] The pair of authors extracts the Lagrangian for a single charge from (2.17). For charge i in an electromagnetic field

$$L_i = -m_i c^2 \sqrt{1 - \frac{\boldsymbol{v}_i^2}{c^2}} + q_i(\boldsymbol{v}_i \boldsymbol{A}^{\mathrm{Lor}} - \Phi^{\mathrm{Lor}}). \tag{8.11}$$

Landau and Lifshitz use the Lorenz gauge and the retarded potentials in their derivation; therefore, Φ^{Lor} and $\boldsymbol{A}^{\mathrm{Lor}}$ are given by (5.1) and (5.2). Both potentials are evaluated at the position $\boldsymbol{r}_i$ of the charge i, and we have also removed the subscript 'ret' to simplify the notation, since in this chapter we only consider the retarded potentials, eliminating any possibility of confusion with the instantaneous potentials, (5.13) and (5.14), and the advanced potentials, (5.22) and (5.23), in the same gauge.

Since the true potentials Φ^{Lor} and $\boldsymbol{A}^{\mathrm{Lor}}$ of the field theory include the action of the charged particles on themselves, the Lagrangian is still formally divergent, $L_i = \infty$, and Landau and Lifshitz replace, *ad hoc*, the true potentials by the external retarded potentials Φ_i^{Lor} and $\boldsymbol{A}_i^{\mathrm{Lor}}$ produced by all the other charges on $\boldsymbol{r}_i$, obtaining

$$L_i = -m_i c^2 \sqrt{1 - \frac{\boldsymbol{v}_i^2}{c^2}} + q_i(\boldsymbol{v}_i \boldsymbol{A}_i^{\mathrm{Lor}} - \Phi_i^{\mathrm{Lor}}). \tag{8.12}$$

In chapters 6 and 7, we used the superscript 'ext', sometimes in subscript form, for the external electromagnetic potentials, forces, and fields, but there is no room for confusion in this chapter since we are dealing with a system of charges and the subscript 'i' allows us to differentiate between the external potentials Φ_i^{Lor} and $\boldsymbol{A}_i^{\mathrm{Lor}}$ acting only on the charge i and the complete potentials Φ^{Lor} and $\boldsymbol{A}^{\mathrm{Lor}}$ that act on all charges simultaneously.

Note: The Lagrangian (8.12) can no longer be interpreted according to the field model of interactions, since the electromagnetic field requires a single set of potentials $(\Phi, \boldsymbol{A})$ common for all the charged particles, and not a set of $2N$ external potentials $(\Phi_i, \boldsymbol{A}_i)$ with a pair of potentials per particle. We can write $\Phi = \sum_i^N \Phi_i + \Phi_i^{\mathrm{self}}$ and $\boldsymbol{A} = \sum_i^N \boldsymbol{A}_i + \boldsymbol{A}_i^{\mathrm{self}}$ in any gauge, but we cannot just remove Φ_i^{self} and $\boldsymbol{A}_i^{\mathrm{self}}$. Physics has to be based on sound mathematics, in which only small quantities are neglected. As Dirac pointed out in his criticism of field theory: *"One is not allowed to neglect infinitely large quantities"*.[44]

In the next step of their heuristic derivation, Landau and Lifshitz expand the external potentials into a power series around instant t, obtaining

$$\Phi_i^{\mathrm{Lor}}(\boldsymbol{r}_i, t) = \sum_{j\neq i}^{N} \int \frac{\kappa}{|\boldsymbol{r}_i - \boldsymbol{r}'|} \left[\rho_j(\boldsymbol{r}', t) + \sum_{s=1}^{\infty} \frac{|\boldsymbol{r}_i - \boldsymbol{r}'|^s}{c^s s!} \frac{\partial^s \rho_j(\boldsymbol{r}', t)}{\partial t^s} \right] \mathrm{d}\boldsymbol{r}' \tag{8.13}$$

and

$$\boldsymbol{A}_i^{\mathrm{Lor}}(\boldsymbol{r}_i, t) = \frac{1}{c^2} \sum_{j\neq i}^{N} \int \frac{\kappa}{|\boldsymbol{r}_i - \boldsymbol{r}'|} \left[\boldsymbol{J}_j(\boldsymbol{r}', t) + \sum_{s=1}^{\infty} \frac{|\boldsymbol{r}_i - \boldsymbol{r}'|^s}{c^s s!} \frac{\partial^s \boldsymbol{J}_j(\boldsymbol{r}', t)}{\partial t^s} \right] \mathrm{d}\boldsymbol{r}', \tag{8.14}$$

where we have used the relation $(t^{\mathrm{ret}} - t) = |\boldsymbol{r} - \boldsymbol{r}'|/c$ in the series expansion.

If we the keep terms up to $(1/c)^2$ order in the power series, use the corresponding charge $\rho_j(\boldsymbol{r}', t) = q_j \delta(\boldsymbol{r}' - \boldsymbol{r}_j(t))$ and current densities $\boldsymbol{J}_j(\boldsymbol{r}', t) = q_j \boldsymbol{v}_j(t) \delta(\boldsymbol{r}' - \boldsymbol{r}_j(t))$ for point charges, expand the free Lagrangian in a power series, keeping the terms up to the same order,

$$-m_i c^2 \sqrt{1 - \frac{\mathbf{v}_i^2}{c^2}} = -m_i c^2 + \frac{m_i \mathbf{v}_i^2}{2} + \frac{m_i \mathbf{v}_i^4}{8c^2} + \mathcal{O}(1/c^4), \tag{8.15}$$

and finally apply a gauge transformation, like Landau and Lifshitz do, we obtain the following Lagrangian for the charge i

$$L_i = \frac{m_i \mathbf{v}_i^2}{2} + \frac{m_i \mathbf{v}_i^4}{8c^2} - \kappa \sum_{j \neq i}^{N} \frac{q_i q_j}{r_{ij}} \left[1 - \frac{\mathbf{v}_i \mathbf{v}_j}{2c^2} - \frac{(\mathbf{v}_i \mathbf{r}_{ij})(\mathbf{v}_j \mathbf{r}_{ij})}{2c^2 r_{ij}^2} \right]. \tag{8.16}$$

From this, the complete Darwin Lagrangian (8.7) is derived by summing the L_i for all the charges and adding an extra (1/2) factor to the interaction part.

Note: Landau and Lifshitz absorb the extra (1/2) factor into a modified summation that includes only one half of the interactions

$$\frac{1}{2} \sum_i^N \sum_{j \neq i}^N = \sum_i^N \sum_{j > i}^N .$$

We can see that to obtain the Darwin Lagrangian, the pair of authors had to include retardation effects up to $(1/c)^2$ order. As we stated before, the interaction Lagrangian (8.10) can be interpreted as the combination of instantaneous and retardation terms. If we had ignored retardation effects in the power series expansion of the potentials (8.13) and (8.14), we would have obtained the alternative result

$$L_{\text{inter}}^{\text{inst}} = -\kappa \sum_i^N \sum_{j \neq i}^N \frac{q_i q_j}{2r_{ij}} \left(1 - \frac{\mathbf{v}_i \mathbf{v}_j}{c^2} \right), \tag{8.17}$$

which is just the instantaneous component of (8.10). Then why do physicists such as Jackson, Schieve, and Trump claim that retardation is neglected in the Darwin Lagrangian? The short answer is that electromagnetic interactions have a dual interpretation, instantaneous and retarded, as we showed in chapter 5.

Landau, Lifshitz, and even Darwin himself use the Lorenz gauge, but others use the Coulomb gauge. In the Coulomb gauge, we can use the pair of instantaneous potentials (5.5) and (5.6) or the mixed pair of an instantaneous scalar potential (5.5) with a retarded transverse vector potential (5.9). Jackson, Schieve, and Trump use the last combination, and we will too in our analysis of their work.

To begin, we must replace the potentials (5.1) and (5.2) for charge i by (5.5) and (5.9) and transform them in external potentials. Again, we will simplify the notation in this chapter by removing unnecessary subscripts and superscripts. The scalar potential is now

$$\Phi_i^{\mathrm{Coul}}(\boldsymbol{r}_i, t) = \sum_{j\neq i}^{N} \int \frac{\kappa}{|\boldsymbol{r}_i - \boldsymbol{r}'|} \rho_j(\boldsymbol{r}', t) \mathrm{d}\boldsymbol{r}', \tag{8.18}$$

while the transverse vector potential, expanded in a power series around the instant t, is

$$\boldsymbol{A}_i^{\mathrm{Coul}}(\boldsymbol{r}_i, t) = \frac{1}{c^2} \sum_{j\neq i}^{N} \int \frac{\kappa}{|\boldsymbol{r}_i - \boldsymbol{r}'|} \left[\boldsymbol{J}_j^{\perp}(\boldsymbol{r}', t) + \sum_{s=1}^{\infty} \frac{|\boldsymbol{r}_i - \boldsymbol{r}'|^s}{c^s s!} \frac{\partial^s \boldsymbol{J}_j^{\perp}(\boldsymbol{r}', t)}{\partial t^s} \right] \mathrm{d}\boldsymbol{r}'. \tag{8.19}$$

To derive the Darwin Lagrangian, Jackson[6] neglects retardation effects in the transverse vector potential, finally using

$$\boldsymbol{A}_i^{\mathrm{Coul}}(\boldsymbol{r}_i, t) = \frac{1}{c^2} \sum_{j\neq i}^{N} \int \frac{\kappa}{|\boldsymbol{r}_i - \boldsymbol{r}'|} \boldsymbol{J}_j^{\perp}(\boldsymbol{r}', t) \mathrm{d}\boldsymbol{r}'. \tag{8.20}$$

This approximation is what Schieve and Trump call the "*magnetostatic limit*".[43] This approximation is what they allude to when they say that the Darwin Lagrangian neglects retardation effects.

Therefore, the interaction term (8.8) can also be interpreted as representing instantaneous interactions between the charges

$$L_{\text{Darw}}^{\text{inter}} = -\kappa \sum_{i}^{N} \sum_{j \neq i}^{N} \underbrace{\frac{q_i q_j}{2r_{ij}} \left[1 - \frac{\boldsymbol{v}_i \boldsymbol{v}_j}{2c^2} - \frac{(\boldsymbol{v}_i \boldsymbol{r}_{ij})(\boldsymbol{v}_j \boldsymbol{r}_{ij})}{2c^2 r_{ij}^2} \right]}_{\text{INSTANTANEOUS}} . \tag{8.21}$$

However, Darwin, Landau, and Lifshitz derive the same Lagrangian by considering retardation effects up to $(1/c)^2$ order, and different gauges must give the same physics. How could the same interaction be both retarded and instantaneous?

The key to this apparent dilemma is that the physical interpretation of the electromagnetic interactions depends on what we chose as source for them. As we showed in chapter 5, using the potentials (5.1) and (5.2) is equivalent to taking retarded ρ_j and $\boldsymbol{J}_j$ as sources of the interactions, while using the potentials (5.5) and (5.9) is equivalent to taking an instantaneous ρ_j and a retarded $\boldsymbol{J}_j^{\perp}$ as the sources.

Of course, we can also have use the instantaneous potentials (5.13) and (5.14) associated with the instantaneous sources Ω_j and $\boldsymbol{S}_j$ or the instantaneous potentials (5.5) and (5.6) associated with the sources ρ_i and $\boldsymbol{J}_i^{\mathrm{T}}$ to derive the interaction term in the Darwin Lagrangian. It is instructive to divide the Maxwell total current $\boldsymbol{J}_i^{\mathrm{T}}$ into conduction and displacement components because if we compare the resulting expression for the interaction Lagrangian term

$$L_{\text{Darw}}^{\text{inter}} = -\kappa \sum_{i}^{N} \sum_{j \neq i}^{N} \underbrace{\frac{q_i q_j}{2r_{ij}} \left[1 - \frac{\boldsymbol{v}_i \boldsymbol{v}_j}{c^2} \right]}_{\text{CONDUCTION}} + \underbrace{\frac{q_i q_j}{2r_{ij}} \left[\frac{\boldsymbol{v}_i \boldsymbol{v}_j}{2c^2} - \frac{(\boldsymbol{v}_i \boldsymbol{r}_{ij})(\boldsymbol{v}_j \boldsymbol{r}_{ij})}{2c^2 r_{ij}^2} \right]}_{\text{DISPLACEMENT}} , \tag{8.22}$$

with the term (8.10), we can observe that the conduction term in (8.22) corresponds to the instantaneous component in (8.10) and the displacement term in (8.22) corresponds to the retardation component in (8.10).

We have found different physical interpretations for the same Lagrangian, depending on whether we use $(\rho_i, \boldsymbol{J}_i)$, $(\rho_i, \boldsymbol{J}_i^{\perp})$, $(\rho_i, \boldsymbol{J}_i^{\mathrm{T}})$, or $(\Omega_i, \mathbf{S}_i)$ as sources of the electromagnetic interactions among charged particles. This explains why some authors consider that the Darwin Lagrangian neglects retardation in the interactions while other authors consider it includes retardation up to $(1/c)^2$ order.

The existence of different interpretations of the same Lagrangian is interesting, but what is really remarkable is that it is impossible for the field theory of electromagnetism to describe the dynamics of a system of interacting charges beyond the $(1/c)^2$ approximation. Landau and Lifshitz write:[7]

> *"We already know that because of the finite velocity of propagation, the field must be considered as an independent system with its own 'degrees of freedom'. From this it follows that if we have a system of interacting particles (charges), then to describe it we must consider the system consisting of these particles and the field. Therefore, when we take into account the finite velocity of propagation of interactions, it is impossible to describe the system of interacting particles rigorously with the aid of a Lagrangian, depending only on the coordinates and velocities of the particles and containing no quantities related to the internal 'degrees of freedom' of the field.*
>
> *However, if the velocity $\mathbf{v}$ of all the particles is small compared with the velocity of light, then the system can be described by a certain approximate Lagrangian. It turns out to be possible to introduce a Lagrangian describing the system, not only when all powers of $\mathbf{v}/c$ are neglected (classical Lagrangian), but also to terms of second order, $\mathbf{v}^2/c^2$."*

Landau and Lifshitz are partially correct in this regard. Indeed, the difficulties to obtain a valid Lagrangian description of the dynamics of N interacting particles appear when we consider $(1/c)^3$ and higher-order terms in the interactions. However, the impossibility of describing the dynamics of a system of interacting charges within the framework of field theory is not exclusive to the use of retarded interactions, because it is also present in the instantaneous formulation.

When the retarded potentials are used to describe interactions, the problem is the existence of an endless causal chain of retarded positions and velocities as in (8.5) and (8.6), while the problem with the instantaneous potentials is an endless sequence of higher-order time derivatives $(\mathrm{d}\mathbf{v}/\mathrm{d}t)$, $(\mathrm{d}^2\mathbf{v}/\mathrm{d}t^2)$, $(\mathrm{d}^3\mathbf{v}/\mathrm{d}t^3)$,..., making it impossible to build a Lagrangian based only on velocities and positions. In both cases we are faced with a description in terms of an infinite number of degrees of freedom.

Another difficulty in electromagnetic field theory appears when we try to define the energy of a system of N interacting charges. As we saw in chapter 4, this theory states that the world is made of charged matter and an electromagnetic field. If this were true, then the energy of matter alone would be a well-defined quantity. The energy of a free charged particle is well defined. Effectively,

$$E^{\mathrm{free}} = \frac{mc^2}{\sqrt{1 - \dfrac{\mathbf{v}^2}{c^2}}}, \tag{8.23}$$

and the energy of N free charges is also well defined and given by

$$E_{\mathrm{N}}^{\mathrm{free}} = \sum_{i}^{N} \frac{m_i c^2}{\sqrt{1 - \dfrac{\mathbf{v}_i^2}{c^2}}}. \tag{8.24}$$

However, what is the total energy of a system of interacting charges? Textbooks on mechanics or electromagnetism usually give the following expression

$$E = \frac{mc^2}{\sqrt{1 - \frac{\mathbf{v}^2}{c^2}}} + e\phi, \tag{8.25}$$

as the energy of a single charge interacting with the electromagnetic field, but even if we ignore the deficiencies of this standard expression (for example that it is formally infinite: $E = \infty$), this expression cannot be used for N charges

$$E_{\mathrm{N}} \neq \sum_{i}^{N} \frac{m_i c^2}{\sqrt{1 - \frac{\mathbf{v}_i^2}{c^2}}} + \sum_{i}^{N} e_i \phi. \tag{8.26}$$

What then is the energy for a system of N-interacting charges? In a section titled "*Relativistic Energy and Momentum*", Griffiths[5] only gives (8.23). The same is true for chapter 11 of Jackson's textbook,[6] and for chapter 4 of Landau and Lifshitz's text.[7] In chapter 5, in a section titled "*Electrostatic energy of charges*", Landau and Lifshitz give the interaction energy for N charges at rest, but throughout the book no expression is given for moving charges. This problem is not unique to field theory, as Feynman and Wheeler never provided an expression for the energy of interacting charges in their alternative action-at-a-distance theory.

Note: Contrary to their claims, Wheeler and Feynman did not provide a fully satisfactory theory of electromagnetism. They simply started from the field approach and integrated out the fields, which did not eliminate all the problems of the field approach. Despite the apparent generality in describing many-body systems, the Wheeler-Feynman theory is "*essentially limited to the case of one particle*".[43] A complete and consistent theory of electromagnetism requires a much deeper analysis of the nature of the interactions.

As the main **conclusions** of this chapter we can offer the following list:

- If field theory cannot provide a complete and satisfactory dynamics for a single body, then logically field theory cannot provide it for two or more bodies.
- The requirement of knowing an infinite number of positions and velocities prohibits the formulation of N-body dynamics as an initial value problem in field theory.
- Since fields are infinite systems, we will never be able to measure or detect them. Also, no one has measured a field locally, but at best physicists and engineers can measure the acceleration of a test charge and then infer the existence of a field.
- It is not true that the Darwin Lagrangian neglects retardation effects. The interaction part of this Lagrangian can be alternatively understood as the sum of an instantaneous term plus retardation corrections up to c^2 order.
- Wheeler and Feynman did not provide a fully satisfactory theory of electromagnetism.

—9

THE COULOMB LIMIT

We mentioned in chapter 3 how Coulomb's original theory was an action-at-a-distance theory, while fields were introduced much latter by Faraday. We also mentioned that:

> *"For a two-particle system, the Coulomb potentials are nonlocal and time-implicit functions $\phi_i(\mathbf{r}_1(t), \mathbf{r}_2(t))$ defined in an abstract configuration space of six dimensions (three per particle), while the potential of the classical field theory is a local and time-explicit function $\Phi(\mathbf{r}, t)$ defined in a four-dimensional spacetime. [...] the potentials of the action-at-a-distance theory are finite and fully physical quantities while the potential of the classical field theory is a divergent quantity that must be corrected by a renormalization procedure."*

Since every conventional textbook on electrodynamics and a large part of the research literature assumes, either explicitly or implicitly, that the Newton-Coulomb theory can be obtained from the Lorentz-Maxwell theory as a limiting case, we will now pay more attention to the relationship between both theories.

Jackson claims[6] that the scalar potential in the Coulomb gauge, the potential (5.5), is "*just the instantaneous Coulomb potential*". To begin with, his wording is misleading because there is no delayed Coulomb potential. The Coulomb potential is instantaneous, and the adjective used by Jackson is redundant. Secondly, (5.5) is not the potential associated to the action-at-a-distance theory developed by Coulomb much before the Maxwell equations were published.

The Coulomb potential for a two-charge system was defined in (3.5). We will now introduce the Coulomb potential for N charges. The potential associated to particle i is

$$\phi_i = \frac{1}{4\pi\epsilon_0} \sum_{j\neq i}^{N} \frac{q_j}{|\boldsymbol{r}_i(t) - \boldsymbol{r}_j(t)|}. \tag{9.1}$$

Using the charge density $\varrho_i(\boldsymbol{r}';t) = \sum_{j\neq i}^{N} q_j \delta(\boldsymbol{r}' - \boldsymbol{r}_j(t))$, we can transform the sum into a space integral

$$\phi_i = \frac{1}{4\pi\epsilon_0} \int \frac{\varrho_i(\boldsymbol{r}';t)}{|\boldsymbol{r}_i(t) - \boldsymbol{r}'|} \mathrm{d}\boldsymbol{r}'. \tag{9.2}$$

On the other hand, the scalar potential (5.5) is, after removing unnecessary subscript and superscript labels,

$$\Phi = \frac{1}{4\pi\epsilon_0} \int \frac{\rho(\boldsymbol{r}',t)}{|\boldsymbol{r} - \boldsymbol{r}'|} \mathrm{d}\boldsymbol{r}'. \tag{9.3}$$

Now we can begin to enumerate the mathematical and physical differences between (9.1) or (9.2) and (9.3); that is, between the potentials of Coulomb theory and field theory.

The first difference between the two theories is that there is a single scalar potential Φ in field theory, while the action-at-a-distance theory of Coulomb assigns a potential ϕ_i to each charged particle.

Therefore, the Coulomb theory is based on a set of N potentials, but this does not mean that it is much more complex than the Maxwell-Lorentz theory, because the potentials in the action-at-a-distance theory are just notational abbreviations, with no physical meaning, since all the physics associated with charge-charge interactions is encoded in the Coulomb potential energy

$$U^{\text{Coul}} = \frac{1}{4\pi\epsilon_0} \sum_i^N \sum_{j\neq i}^N \frac{q_i q_j}{|\boldsymbol{r}_i(t) - \boldsymbol{r}_j(t)|}. \tag{9.4}$$

There are N potentials in the Coulomb theory, because a charge cannot interact with itself by virtue of the condition $j \neq i$ in the summations. This means that another difference is that the Coulomb potentials ϕ_i are finite, while the scalar potential diverges, $\Phi = \infty$. We will go into more details in appendix A.

A third difference is that the Coulomb potentials are *explicit* functions of the positions of the N charges and *implicit* functions of time $\phi_i = \phi_i(\boldsymbol{r}_1(t), \boldsymbol{r}_2(t), \ldots \boldsymbol{r}_N(t))$, while the scalar potential is an *explicit* function of space-time variables $\Phi = \Phi(\boldsymbol{r}, t)$. The Coulomb potentials are defined in a $3N$-dimensional configuration space, while the scalar potential of the field theory is defined in a 4-dimensional Minkowski space.

I have mentioned Jackson, but similar erroneous material can be found in the texts of other authors including Landau and Lifshitz.[7] Furthermore, since the Coulomb potential of the action-at-a-distance theory does not depend explicitly on time, some authors identify this potential with the electrostatic limit of (9.3); that is, with the solution $\Phi = \Phi(\boldsymbol{r})$ to the Poisson equation

$$\Phi = \frac{1}{4\pi\epsilon_0} \int \frac{\rho(\boldsymbol{r}')}{|\boldsymbol{r} - \boldsymbol{r}'|} \mathrm{d}\boldsymbol{r}'. \tag{9.5}$$

This is the case with Greiner[8] or with Griffiths.[5,19] We can even find texts that contradict themselves, as is the case with Schwinger et al.,[9] who refer to (9.3) as the "*Coulomb potential*", but about ten chapter before, they refer to (9.5) as the "*well-known Coulomb potential*". In any case, neither $\Phi(\boldsymbol{r}, t)$ nor $\Phi(\boldsymbol{r})$ are the potential of the Coulomb theory.

The confusion between the potentials ϕ_i of the Coulomb theory and the scalar potential Φ of the Maxwell-Lorentz theory is ubiquitous in mainstream electrodynamics textbooks and in much of the research literature, but some exceptions can be found. One of them is the monograph[43] by Trump and Schieve, another is a series of papers[45,46] by Chubykalo and Smirnov-Rueda (see also the comment[47] by Ivezić and Škovrlj and the corresponding reply[48]).

Trump and Schieve are motivated by the development of a complete and consistent many-body theory valid for any speed. They start from the Coulomb potential (9.1) and try to find a high-velocity version without the problems of the field theory. Their proposal is

$$\phi_i^{\mathrm{SHP}} = \frac{1}{4\pi\epsilon_0} \sum_{j\neq i}^{N} \frac{q_j}{|\boldsymbol{\chi}_i(\tau) - \boldsymbol{\chi}_j(\tau)|}. \tag{9.6}$$

where $\boldsymbol{\chi}_i(\tau)$ are Minkowski-like four-distances with τ an invariant evolution parameter for the N-body system, and the superscript 'SHP' refers to the theory that Trump and Schieve use.

Note: The acronym SHP is common in the literature and refers to Stueckelberg, Horwitz, and Piron.

The Coulomb potentials are defined in a $3N$-dimensional configuration space, while the SHP potentials are defined in a $4N$-dimensional configuration space. For a two-body system we go from six degrees of freedom to eight.

The SHP potentials produce all kinds of problems and inconsistencies. This is a direct consequence of the uncritical mixing of theoretical elements from the action-at-a-distance theories developed by Newton and Coulomb with elements from special relativity. Trump and Schieve correctly criticize the flaws of the so-called relativistic field theory (see the quote on page 98), but they fail to mention that some of those flaws were inherited by special relativity, since the latter was derived from the Maxwell-Lorentz theory. It can be shown that the approach followed by Trump and Schieve is not necessary to develop a consistent many-body theory for high velocities. I have only cited Trump and Schieve here because they give the correct form for the Coulomb potential, unlike Jackson, Landau, Lifshitz, Greiner, Griffiths, Schwinger, and many other mainstream authors.

Let us now see what Chubykalo and Smirnov-Rueda have to say about the Coulomb potential and the field theory. They begin with a mathematical analysis of the potential form of the Maxwell equations in the Lorenz gauge, equations (2.11) and (2.12), and write:[45]

> *"Differential equations have, generally speaking, an infinite number of solutions. A uniquely determined solution is selected by laying down sufficient additional conditions. Different forms of additional conditions are possible for the second-order partial differential equations: initial value and boundary conditions. Usually, a general solution of D'Alembert's equation is considered as an explicit time-dependent function $g(\mathbf{r}, t)$. In the stationary state the D'Alembert equation is transformed into the Poisson equation, whose solution is an implicit time-dependent function $f(\mathbf{R}(t))$. Nevertheless, the conventional theory does not explain in detail how the function $g(\mathbf{r}, t)$ is converted into an implicit time-dependent function $f(\mathbf{R}(t))$ (and vice versa) when the steady-state problems are studied."*

Then Chubykalo and Smirnov-Rueda propose combining both types of solutions in a linear way. In the case of the Coulomb potential and the scalar potential, the combination that they suggest is

$$\Phi_0(\boldsymbol{R}(t)) + \Phi(\boldsymbol{r}, t). \tag{9.7}$$

However, this linear combination is not valid for several reasons. First, the functional form $\Phi_0(\boldsymbol{R}(t))$, with a single charge-charge distance, is only valid for two-body interactions, because for N charges there are $N(N-1)/2$ distances. Second, the Coulomb theory requires N potentials, one for each charge, and not a single potential that can be combined with the scalar potential of the field theory.

Furthermore, the claim[45] that the "*electrodynamics dualism concept*" must be introduced into physics because "*there is a simultaneous and independent coexistence of Newton instantaneous long-range (NILI) and Faraday-Maxwell short-range interactions (FMSI), which cannot be reduced to each other*" is not true. We show in appendix A how $\Phi(\boldsymbol{r}, t)$ can be derived from the Coulomb potentials as a local approximation. I have only cited the work of Chubykalo and Smirnov-Rueda here because they do not completely confuse the potentials of the theory of Coulomb with the potentials of the field theory, not because I agree with their dualistic approach to electromagnetic interactions.

Confusing the potential of the action-at-a-distance theory developed by Coulomb with the instantaneous scalar potential of the field theory is not the only error that we can find in the literature; we find related misconceptions in the mainstream textbooks on electrodynamics and a large part of the research literature. For example, the Coulomb law is often expressed as[5]

$$\boldsymbol{F}(\boldsymbol{r}) = q\boldsymbol{E}(\boldsymbol{r}), \tag{9.8}$$

when the true Coulomb law, an action-at-a-distance law, can be derived from the potential energy (9.4) for a system of N charged particles by the usual method: $\boldsymbol{F}^{\mathrm{Coul}} = -\nabla U^{\mathrm{Coul}}$. For charge i

$$\boldsymbol{F}_i^{\mathrm{Coul}}(\boldsymbol{r}_1(t), \boldsymbol{r}_2(t), \ldots \boldsymbol{r}_N(t)) = \frac{1}{4\pi\epsilon_0} \sum_{i}^{N} \sum_{j \neq i}^{N} q_i q_j \frac{\boldsymbol{r}_i(t) - \boldsymbol{r}_j(t)}{|\boldsymbol{r}_i(t) - \boldsymbol{r}_j(t)|^3}, \quad (9.9)$$

which differs from the field-theoretic expression (9.8) both physically and mathematically.

The main **conclusions** of this chapter are the following:

- The Coulomb theory cannot be considered a limiting case of the Lorentz-Maxwell theory.
- The scalar potential of the field theory in the Coulomb gauge is not an "*instantaneous Coulomb potential*", but a completely different potential, both mathematically and physically.
- In field theory, there is a single scalar potential for a system of N interacting charges, while the Coulomb theory is based on a set of N Coulomb potentials, one per charged particle.
- The Coulomb potentials are explicit functions of the positions of the N charges and implicit functions of time, while the scalar potential of field theory is an explicit function of spatio-temporal variables. Consequently, the Coulomb potentials are defined in a $3N$-dimensional configuration space, while the scalar potential is defined in a 4-dimensional Minkowski spacetime.
- The Coulomb force law of the theory of action at a distance is also confused with the force derived in the Lorentz-Maxwell theory from the electric component of the field.

—10

FLUCTUATIONS

Electrodynamics textbooks take the Maxwell-Lorentz equations as if they were always valid for any system where gravitation and quantum effects are negligible. Physicists like Jackson mention the existence of a set of "*microscopic Maxwell equations*" that govern electromagnetic phenomena in the "*microscopic world made up of electrons and nuclei*".[6] However, the Maxwell-Lorentz equations are only valid for macroscopic systems in a mean sense; that is, when fluctuations due to the discrete nature of matter are ignored.

To mechanically describe any classical macroscopic system, we must know the position and velocity (or momentum) of each particle, plus additional quantities such as mass and electric charge. In a cubic crystalline region with a side length of 0.01 centimeters, we can find about 10^8 charged particles, which implies about 10^9 classical variables. Since there is no way to keep track of all the variables, measurements are limited to the specification of a small number of characteristic quantities, the *reduced description*, which is many orders of magnitude smaller than the number of variables necessary to provide a complete description of the system.

It is clear that two or more systems that have the same contracted description do not have to be identical in all respects. This follows from the fact that several complete descriptions can result in the same reduced description. As a consequence, the reduced description has built-in fluctuations. Fluctuations are (typically) small and unpredictable variations in a given quantity. Fluctuations are sometimes called noise, especially in engineering contexts.

To deal with the differences between systems with identical contracted descriptions, it is natural to introduce the idea of a physical ensemble. A physical ensemble consists of a collection of systems identically prepared with respect to their contracted description.

Note: The concept of physical ensemble is related to the concept of statistical ensemble introduced by Gibbs into classical statistical mechanics, but would not be confused with it. Gibbs ensembles are described by a probability density function in phase space, while a physical ensemble is characterized by a reduced description.

Since the reduced description of any macroscopic system incorporates fluctuations, the natural formal framework for dealing with physical ensembles is the theory of stochastic quantities and stochastic processes. Stochastic quantities are physical variables that include fluctuations. The existence of fluctuations means that there is a distribution of possible values of the quantity, any of which could be found in an actual measurement on a system of the physical ensemble.

Suppose that we have a collection of g systems with an identical contracted description, with g being a very large number. If we simultaneously measure the same reduced variable Z in all the systems in the collection, we will typically find small discrepancies in the recorded values. After making g measurements, we have a list of Z_k values, one for each system k, and we can apply basic statistical formulae to this list. We define the average Z of the reduced variable at time t as

$$Z(t) = \frac{1}{g} \sum_{k=1}^{g} Z_k(t). \tag{10.1}$$

The value of this variable in the system k at time t is $Z_k(t)$, while $Z(t)$ represents the value associated with the physical ensemble. In general, both values differ and we can write

$$Z_k(t) = Z(t) + \delta Z_k(t). \tag{10.2}$$

The quantity δZ_k is what we call fluctuation, and its value can be positive, negative, or zero. We can say that the value of the property in a given system k is equal to a value Z that is common to all the systems in the physical ensemble plus a deviation δZ_k that characterizes this system.

We are generally not interested in collections of identically prepared systems, because we only prepare and study one system at a time. In this case, we can simplify the notation by eliminating the subscript k, since the measurements will always refer to the same system, but to avoid confusion between the value of the property in the system and the value in the associated ensemble, we will use the tilde ~ for the value in the system. With this new notation, the expression

$$\tilde{Z}(t) = Z(t) + \delta Z(t) \tag{10.3}$$

denotes that the measured variable $\tilde{Z}$ in the system under study can be characterized as the sum of an average value Z plus a fluctuation δZ.

Note: The value of the property that we measure in experiments is $\tilde{Z}$. We never measure Z in a system except in the special case when the fluctuation is zero, which is an abnormally unusual situation. The value Z is obtained through statistical analysis of the values obtained experimentally.

In the previous chapters, we considered only average values for the electromagnetic quantities because textbooks on electrodynamics ignore fluctuations. However, our measurements of electromagnetic quantities include fluctuations. Figure 10.1 compares the average electric strength described by the ordinary Maxwell equations in the electrostatic regime with the fluctuating field associated to the stochastic motion of the charges.

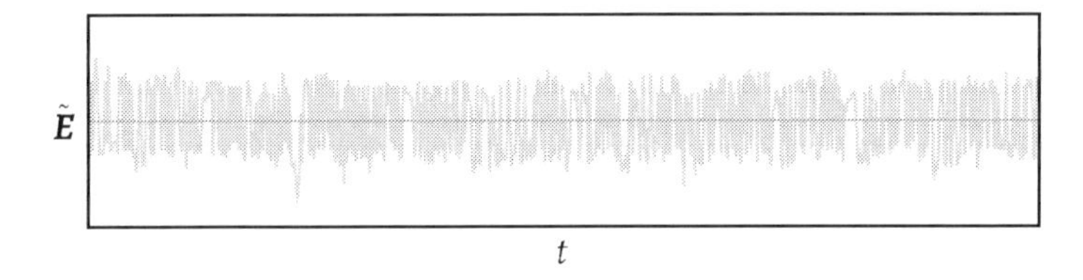

Figure 10.1: Random variations of the electric strength with time. The horizontal line represents the average value of the component $\boldsymbol{E}$ of the electromagnetic field.

In this chapter, we will consider the modifications that must be made to the usual electromagnetic theory in order to include fluctuations. Replacing the average values of the mechanical and electromagnetic quantities in the equations found in textbooks with the experimental stochastic values of these quantities is not enough to correctly describe real world electromagnetic phenomena. It is also necessary to add additional terms to some of those equations.

Consider the ordinary Lorentz equation (6.11), not only do we have to replace $\boldsymbol{E}$ with $\tilde{\boldsymbol{E}}$, $\boldsymbol{v}$ with $\tilde{\boldsymbol{v}}$, and so on, but we must also include an extra force term. The stochastic Lorentz equation is

$$\frac{\mathrm{d}(m\tilde{\gamma}\tilde{\boldsymbol{v}})}{\mathrm{d}t} = q(\tilde{\boldsymbol{E}} + \tilde{\boldsymbol{v}} \times \tilde{\boldsymbol{B}}) + \tilde{\boldsymbol{F}}^{\mathrm{rand}}, \tag{10.4}$$

with $\tilde{\boldsymbol{F}}^{\text{rand}}$ an additional random force that does *not* have the form of the stochastic Lorentz force.

Similarly, the stochastic Maxwell equations are[49,50]

$$\nabla \cdot \tilde{\boldsymbol{E}} = \frac{\tilde{\rho}}{\epsilon_0}, \tag{10.5}$$

$$\nabla \cdot \tilde{\boldsymbol{B}} = 0, \tag{10.6}$$

$$\nabla \times \tilde{\boldsymbol{E}} + \frac{\partial \tilde{\boldsymbol{B}}}{\partial t} = \frac{1}{\epsilon_0 c} \tilde{\boldsymbol{J}}_{\text{e}}^{\text{rand}}, \tag{10.7}$$

and

$$\nabla \times \tilde{\boldsymbol{B}} - \frac{1}{c^2} \frac{\partial \tilde{\boldsymbol{E}}}{\partial t} = \mu_0 \tilde{\boldsymbol{J}} + \mu_0 \tilde{\boldsymbol{J}}^{\text{rand}}, \tag{10.8}$$

with $\tilde{\boldsymbol{J}}_{\text{e}}^{\text{rand}}$ and $\tilde{\boldsymbol{J}}^{\text{rand}}$ two random currents. We recover the ordinary Maxwell equations found in textbooks when we take averages on both sides of equations (10.5), (10.6), (10.7), and (10.8), since the averages of the random currents vanish: $\boldsymbol{J}_{\text{e}}^{\text{rand}} = \boldsymbol{J}^{\text{rand}} = \boldsymbol{0}$. However, the average of the random force in (10.4) cannot be zero in general, because this equation is not linear in the fluctuations.

Moreover, we can generalize (10.4) by including other stochastic forces; for example, thermal forces due to the interactions with a heat bath. A popular choice is the Langevin-Lorentz equation

$$\frac{\mathrm{d}(m\tilde{\gamma}\tilde{\boldsymbol{v}})}{\mathrm{d}t} = q(\tilde{\boldsymbol{E}} + \tilde{\boldsymbol{v}} \times \tilde{\boldsymbol{B}}) - \zeta\tilde{\gamma}\tilde{\boldsymbol{v}} + \tilde{\boldsymbol{F}}^{\text{rand}}, \tag{10.9}$$

where ζ is the friction coefficient for the charge-bath coupling. As we already know, the left-hand side of this equation of motion can be replaced with $m\tilde{\boldsymbol{a}}$ in the low-velocity limit, since $\tilde{\gamma} \approx 1$, and the resulting equation has the following form

$$m\tilde{\boldsymbol{a}} = q(\tilde{\boldsymbol{E}} + \tilde{\boldsymbol{v}} \times \tilde{\boldsymbol{B}}) - \zeta\tilde{\boldsymbol{v}} + \tilde{\boldsymbol{F}}^{\text{rand}}. \tag{10.10}$$

This low-velocity limit of the Langevin-Lorentz equation is used in the research literature for applications to physical and biological systems.[51–53] You will not find this useful equation of motion in ordinary textbooks.[4–9]

The main **conclusions** of this chapter are the following:

- Contrary to myth, the Maxwell-Lorentz equations are not valid for "*any system where gravitation and quantum effects are negligible*".
- The reduced description of any macroscopic system incorporates fluctuations.
- The stochastic Lorentz equation contains an additional random force that is not in the form of a stochastic Lorentz force.
- The Maxwell equations found in electrodynamics textbooks correspond to the average of the stochastic Maxwell equations.

—11

COVARIANT FORMULATION

Until now we have used the traditional formulation of electrodynamics in terms of scalars and three-dimensional vectors. However, the development of special relativity introduced a four-dimensional view of mechanics and this prompted physicists to develop a new formulation of electrodynamics. This is the "*covariant formulation of electrodynamics*",[8] in which related scalars and three-dimensional vectors are combined into four-dimensional vectors and higher-order tensors, as we will see in a moment.

Position and time are combined into a so-called four-position vector $r^\alpha = (ct, \boldsymbol{r})$, where the index α runs over the four values 0, 1, 2, and 3, so that $r^0 = ct$ and r^1, r^2, and r^3 are the components of $\boldsymbol{r}$. Differentiating the four-position r^α with respect to the proper time τ we obtain the four-velocity

$$u^\alpha = \frac{\mathrm{d}r^\alpha}{\mathrm{d}\tau} = (u^0, \boldsymbol{u}) = (\gamma c, \gamma \mathbf{v}), \tag{11.1}$$

where $\gamma = (\mathrm{d}t/\mathrm{d}\tau)$ is again the time-dilation factor.

The four-gradient is defined as $\partial^\alpha = (\partial/c\partial t, -\nabla)$ and a four-momentum $p^\alpha = (E/c, \boldsymbol{p})$ can be obtained by combining momentum and energy. Electric charge and current densities are combined into a four-current density $J^\alpha = (c\rho, \boldsymbol{J})$ and a four-potential $A^\alpha = (\phi, \boldsymbol{A})$ is obtained from the scalar and vector potentials.

Note: The covariant notation we are using is standard‡ but ambiguous, since the same symbol, for example p^α, is used to denote both the four vector and one of its components. This ambiguity is not present in the noncovariant formulation, in which we use $\boldsymbol{p}$ for the momentum and p^i for one of its components, so that $\boldsymbol{p} = (p^1, p^2, p^i)$. The term four-velocity is also a standard name for u^α, but it is misleading, because the spatial component $\boldsymbol{u}$ is not a velocity, but a proper velocity. Authors like Charap[54] pretend that it is better to call $\boldsymbol{v}$ a three-velocity, which should mess up the terminology even more.

‡ The only difference is that we use $\boldsymbol{r} = (x, y, z)$ in this book to denote position, and so we use r^α for the four-position. Electrodynamics and special relativity textbooks use $\boldsymbol{x}$ for position and x^α for the four-position, which increases the ambiguity of the notation since x is also one of the components of position.

$\boldsymbol{E}$ and $\boldsymbol{B}$ are three-dimensional vectors and cannot be combined into a four-vector, but they can be combined into a higher-order tensor: the electromagnetic field tensor $F^{\alpha\beta}$

$$F^{\alpha\beta} = \begin{pmatrix} 0 & -E_x/c & -E_y/c & -E_z/c \\ E_x/c & 0 & -B_z & B_y \\ E_y/c & B_z & 0 & -B_x \\ E_z/c & -B_y & B_x & 0 \end{pmatrix}. \tag{11.2}$$

To develop the covariant formulation, it is also necessary to introduce the Minkowski metric tensor

$$\eta^{\alpha\beta} = \begin{pmatrix} 1 & 0 & 0 & 0 \\ 0 & -1 & 0 & 0 \\ 0 & 0 & -1 & 0 \\ 0 & 0 & 0 & -1 \end{pmatrix}. \tag{11.3}$$

Note: The diagonal of the Minkowski metric tensor is $(+1, -1, -1, -1)$ because I am using the particle physicists' convention, also called the timelike or West-coast convention, for the metric signature. There is another choice with the signs swapped, $(-1, +1, +1, +1)$ this is the convention used by general relativists, also called the spacelike or East-coast convention, but the physics described by both conventions is the same.

The metric tensor can be used to define a new set of quantities, in a process known informally as the lowering of indices, for reasons that will become evident in a moment. First, we set $\eta_{\alpha\beta} = \eta^{\alpha\beta}$ and then use this to define new vectors like

$$A_\beta = \sum_{\alpha=0}^{3} \eta_{\alpha\beta} A^\alpha \tag{11.4}$$

and new higher-order tensors like

$$F_{\lambda\zeta} = \sum_{\alpha=0}^{3} \sum_{\beta=0}^{3} \eta_{\alpha\lambda} \eta_{\zeta\beta} F^{\alpha\beta}. \tag{11.5}$$

The sums in the above expressions are often not written explicitly. This is the so-called *Einstein notation*, a convention in which repeating indices are implicitly summed over, for brevity of notation. With this convention, we write $A_\beta = \eta_{\alpha\beta} A^\alpha$ and $F_{\lambda\zeta} = \eta_{\alpha\lambda}\eta_{\zeta\beta} F^{\alpha\beta}$, respectively. We will use the Einstein notation in the rest of this chapter.

The Minkowski spacetime line element in a covariant formulation is given by

$$\mathrm{d}s^2 = \mathrm{d}x^\alpha \eta_{\alpha\beta} \mathrm{d}x^\beta, \tag{11.6}$$

and proper time is

$$\mathrm{d}\tau = \frac{1}{c}\sqrt{\mathrm{d}x^\alpha \eta_{\alpha\beta} \mathrm{d}x^\beta}. \tag{11.7}$$

With this basic machinery, we can begin to reformulate electrodynamics and mechanics.

The Lorentz equation (6.19) is replaced by the four-dimensional equation

$$\frac{\mathrm{d}(mu^{\alpha})}{\mathrm{d}\tau} = K^{\alpha}, \tag{11.8}$$

where $K^{\alpha} = qF^{\alpha\beta}u_{\beta}$ is the so-called "*electromagnetic Minkowski force*", another unfortunate name, since K^{α} is actually a four-force and only $\boldsymbol{K} = (K^1, K^2, K^3) = \gamma\boldsymbol{F}^{\mathrm{Lor}}$ is a force. Expression (11.8) is the Minkowski-Lorentz equation.

The law of conservation of charge (2.5) takes the covariant form

$$\partial_{\alpha}J^{\alpha} = 0. \tag{11.9}$$

The Maxwell equations (2.1) and (2.4) can be combined into a single equation

$$\partial_{\beta}F^{\alpha\beta} = \mu_0 J^{\alpha}, \tag{11.10}$$

but to rewrite the other pair of Maxwell equations in a similar way, we need to introduce[5] a *dual electromagnetic tensor* $G^{\alpha\beta}$, also called the dual field strength tensor,[6,54] which can be obtained directly from $F^{\alpha\beta}$ by the substitution $\mathbf{E}/c \to \mathbf{B}$ and $\mathbf{B} \to -\mathbf{E}/c$,

$$G^{\alpha\beta} = \begin{pmatrix} 0 & -B_x & -B_y & -B_z \\ B_x & 0 & E_z/c & -E_y/c \\ B_y & -E_z/c & 0 & E_x/c \\ B_z & E_y/c & -E_x/c & 0 \end{pmatrix}. \tag{11.11}$$

With this new tensor the Maxwell equations (2.2) and (2.3) can be combined into a single equation

$$\partial_{\beta}G^{\alpha\beta} = 0. \tag{11.12}$$

Some authors[8] do not use the dual tensor, and prefer to combine (2.2) and (2.3) as

$$\partial_{\zeta}F^{\alpha\beta} + \partial_{\alpha}F^{\beta\zeta} + \partial_{\beta}F^{\zeta\alpha} = 0, \tag{11.13}$$

but this is a matter of taste.

Finally, the Maxwell equations in the potential formulation, equations (2.8) and (2.9) can be written in covariant form as

$$\partial_\beta \partial^\beta A^\alpha = -\mu_0 J^\alpha. \tag{11.14}$$

A priori, it seems we have achieved a high degree of compactness with the new formulation since, for example, we have reduced the original four Maxwell equations to only two: (11.10) and (11.12). However, this compactness is only apparent, because the original equations use the vectors $\boldsymbol{E}$ and $\boldsymbol{B}$, with three components each, while the covariant notation requires a pair of 4×4 tensors: $F^{\alpha\beta}$ and $G^{\alpha\beta}$. Similarly, the Lorentz equation is replaced in this new formulation by a four-dimensional equation (11.8) that contains redundancies.

Before analyzing the redundancies in (11.8), we will consider a simpler mechanical expression. For a free particle, its energy is $E = \sqrt{m^2c^4 + \mathbf{p}^2c^2}$, which implies that $m^2c^4 = E^2 - \mathbf{p}^2c^2$ or, in covariant form,

$$m^2c^4 = p^\alpha \eta_{\alpha\beta} p^\beta. \tag{11.15}$$

This relationship between mass and four-momentum is the *mass-shell condition* and means that the four-momentum p^α is not a true four-vector in the sense that its components (p^0, p^1, p^2, p^3) cannot vary freely. For example, if we give values to p^1, p^2, and p^3 then (11.15) fixes the value of p^0 for a given mass. The covariant formulation is mathematically four-dimensional, but the physics is only three-dimensional and this gets reflected in the apparition of redundant degrees of freedom in the covariant quantities and equations.

Note: To construct true four-dimensional formulations, some physicists ignore the mass-shell condition, allowing the components of the four-momentum to take arbitrary values.[43] These fully covariant formulations are purely formal and to make contact with reality the condition must ultimately be imposed.

For an interacting particle, the mass-shell condition changes to

$$m^2c^2 = (p^\alpha - qA^\alpha)\,\eta_{\alpha\beta}\,(p^\beta - qA^\beta) \tag{11.16}$$

and using the relationship $(p^\alpha - qA^\alpha) = mu^\alpha$, and differentiating with respect to proper time, we obtain the following constraint

$$0 = \frac{\mathrm{d}(mu^\alpha)}{\mathrm{d}\tau}\eta_{\alpha\beta}u^\beta, \tag{11.17}$$

which helps us to understand why (11.8) is a not a true four-dimensional equation of motion, since the magnitude of the four-velocity for any object is always a fixed constant. The Minkowski-Lorentz equation is only mathematically four-dimensional. The motion of the charged particle is physically determined in three dimensions and once we solve the Lorentz equation we can obtain the value of $\mathrm{d}(mu^0)/\mathrm{d}\tau$ from the above constrain.

With this brief introduction to the covariant formulation of electrodynamics and mechanics, we are ready to begin reviewing the main misconceptions found in the literature. Of course, we will review only the new misconceptions that arise with the new formulation and ignore the misconceptions that we discussed in the previous chapters. So that, the Minkowski-Lorentz equation (11.8) has essentially the same difficulties as the Lorentz equation (6.19) with respect to energy conservation, causation, divergences, and inertia. Similarly, if we replace the Lagrangian (2.17) with a covariant action, such as Landau and Lifshitz do,[7] we will not eliminate the difficulties associated with the physical interpretation of each term.

This is quite obvious since a new formulation does not change the physics and if this physics is incorrect, limited to special cases, or has problems of interpretation, then the difficulties remain in the new formulation. For the same reason, the ordinary claim that the quantity mu^α is "*the only reasonable candidate for a spacetime tensor that could represent the momentum of the particle*"[10] is incorrect, because its spatial component, the term $m\gamma\mathbf{v}$, only represents the momentum of the particle when this is free, how we saw in previous chapters.

Note: That mu^{α} is not the four-momentum of the particle is the reason why the four-momentum we must quantize in a covariant formulation of quantum theory is $p^{\alpha} = (mu^{\alpha} + qA^{\alpha})$, which becomes the four-operator $\hat{p}^{\alpha} = i\hbar\partial^{\alpha}$.

But before reviewing the new misconceptions associated to the covariant formulation, I want to point out that Griffiths[5] refers to this reformulation as "*relativistic electrodynamics*" even though all the formulas presented in the previous chapters are already consistent with special relativity. It should be emphasized, however, that Griffiths correctly states that the new formulation does not change "*the rules of electrodynamics in the slightest*", but rather the new formulation simply expresses these rules "*in a notation that exposes and illuminates their relativistic character*". Despite this useful qualification, the term "*relativistic electrodynamics*" remains a misnomer.

Jackson claims that "*the mathematical equations expressing the laws of nature must be covariant, that is, invariant in form, under the transformations of the Lorentz group*".[6] Wald goes further in his textbook[10] and states that expressions like the Lorentz equation (6.19) involving "*spatial vectors have no well-defined status in special relativity, nor does t*", and so "*it must be discarded and replaced by a new law that involves only spacetime tensors*" like (11.8). This is a typical misconception.

The principle of relativity states that the observations of different observers must be mutually compatible, and not that they must observe the same. Equations like (6.19) are fully relativistic and valid for any inertial observer (the only observers considered by special relativity). Four-dimensional equations like (11.8) are only useful because it is *simpler* to verify that they transform correctly under the Lorentz group transformations associated to special relativity. However, this advantage comes at the expense of redundancies, as discussed above.

To show that three-dimensional equations are fully compatible with the principle of relativity, consider two inertial observers, one of them at rest and using the reference frame $r^\alpha = (ct, \boldsymbol{r})$, and the other observer using $r'^\alpha = (ct', \boldsymbol{r}')$ and moving with a velocity $\boldsymbol{w}$ with respect to the first. Since we can obtain the Lorentz equation (6.19) from the spatial component of the Minkowski-Lorentz equation (11.8), we can write the following diagram relating the equations of motion used by each observer to analyze and predict phenomena

$$\begin{array}{ccc} \dfrac{\mathrm{d}(mu^\alpha)}{\mathrm{d}\tau} = K^\alpha & \xleftrightarrow{\text{inertial transformation}} & \dfrac{\mathrm{d}(mu'^\alpha)}{\mathrm{d}\tau'} = K'^\alpha \\ \Big\downarrow \text{spatial projection} & & \text{spatial projection} \Big\downarrow \\ \dfrac{\mathrm{d}(m\gamma\boldsymbol{v})}{\mathrm{d}t} = \boldsymbol{F}^{\mathrm{Lor}} & \xleftrightarrow[\text{inertial transformation}]{} & \dfrac{\mathrm{d}(m\gamma'\boldsymbol{v}')}{\mathrm{d}t'} = \boldsymbol{F}'^{\mathrm{Lor}} . \end{array} \tag{11.18}$$

We can now see how explicitly covariant equations simplify the task of transforming the observations of one of the inertial observers into those of the other, in such a way that their respective observations are mutually consistent. This simplification can be observed in the variable time since both observers can use the same proper time, $\mathrm{d}\tau = \mathrm{d}\tau'$, but they cannot use the same time since $\mathrm{d}t \neq \mathrm{d}t'$, which implies that before comparing their observations, the times they measured with their respective clocks must be Lorentz transformed $t \longleftrightarrow t'$. Furthermore, the Minkowski four-forces for each observer are related simply by a Lorentz transformation $qF^{\alpha\beta}u_\beta \longleftrightarrow qF'^{\alpha\beta}u'_\beta$, unlike the Lorentz forces, which require a more complex transformation.

There is another difficulty with the Minkowski-Lorentz equation (11.8). This equation uses proper time, which is the time measured by a hypothetical clock attached to the charged particle. What we measure in the laboratory is t not τ. Of course, the reader might object that both times are related by $\gamma = (\mathrm{d}t/\mathrm{d}\tau)$ and therefore using proper time does not really introduce any practical difficulty. However, things become more complex when we want to study the motion of N charged particles, since each particle will have its own proper time. In the many-body case, equation (11.8) would be replaced by a set of N equations as

$$\frac{\mathrm{d}(m_i u_i^{\alpha})}{\mathrm{d}\tau_i} = K_i^{\alpha}. \tag{11.19}$$

As we can see, not only does the explicitly covariant formalism increase complexity by introducing redundant components associated with four-vectors and higher-order tensors, but for the many-body case the covariant equations of motion require N times to describe the dynamics. This fact implies that we cannot combine all the N covariant equations of motion into a single equation like we can do in ordinary mechanics, since we no longer have a common time for all particles in the covariant formulation. In fact, if we take a look at the aforementioned texts by Jackson and Wald, we can see that they never write a N-body equation of motion, but only individual equations for each charged particle as if the particles were moving *"in external electromagnetic fields (neglecting the emission of radiation)"*.[6]

Another set of difficulties with the new formulation is related to the choice of explicitly covariant Lagrangians and Hamiltonians. The Lagrangian L for a charged particle on an electromagnetic field is (8.11). Using a Legendre transformation, we can derive the corresponding Hamiltonian (6.2) from that Lagrangian. The Lagrangian (8.11) and the Hamiltonian (6.2) are what we use in practical applications, but they are not invariant to Lorentz transformations. So if an inertial observer at rest is using the Lagrangian L to derive the equations of motion in his reference frame, then another inertial observer moving with respect to the first one, will have to use a $L' \neq L$ to derive the corresponding equations of motion. For this reason, many physicists pretend we would only use covariant Lagrangians.

An explicitly covariant Lagrangian is[8,10,54]

$$\mathfrak{L} = -mc^2 + qu^\alpha A_\alpha. \tag{11.20}$$

This expression is explicitly covariant since the first term is minus a rest energy and the second term is the scalar product of two four-vectors. $\mathfrak{L}$ is a Lorentz invariant, which means that $\mathfrak{L} = \mathfrak{L}'$ for any two inertial observers. This seems like an advantage. However, we cannot use (11.20) to derive the equations of motion since the first term is not a explicit function of the four-velocities. We have to rewrite this Lagrangian by using the mass-shell condition (11.16) in the form $c^2 = u^\alpha \eta_{\alpha\beta} u^\beta$. However, we cannot use this one either but have to use its square root, finally obtaining the alternative expression[6]

$$\mathfrak{L} = -mc\sqrt{u^\alpha \eta_{\alpha\beta} u^\beta} + qu^\alpha A_\alpha. \tag{11.21}$$

By introducing this expression in a covariant version of the Euler-Lagrange equations, we can obtain the covariant equation of motion (11.8), with all the mentioned advantages and disadvantages, but now the fun begins. Using a covariant version of the Legendre transformation, we can derive the following covariant Hamiltonian[6] from (11.21)

$$\mathfrak{H} = \frac{(p^\alpha - qA^\alpha)\,\eta_{\alpha\beta}\,(p^\beta - qA^\beta)}{m} - c\sqrt{(p^\alpha - qA^\alpha)\,\eta_{\alpha\beta}\,(p^\beta - qA^\beta)} \tag{11.22}$$

This is a Lorentz invariant, but it is not a true Hamiltonian since, using again (11.16), we can show that $\mathfrak{H} = 0$ and we know that the energy of an electromagnetic system is not constrained to zero.

We will call quantities like this a pseudo-Hamiltonian. Jackson writes:

> *"While the Hamiltonian above is formally satisfactory, it has several problems. The first is that it is by definition a Lorentz scalar, not an energylike quantity. Second, use of [...] shows that $\mathfrak{H} = 0$. Clearly, such a Hamiltonian formulation differs considerably from the familiar nonrelativistic version."*

Jackson then cites a reference[55] by Barut for an extended discussion of $\mathfrak{H}$ and other pseudo-Hamiltonians. However, Jackson fails to mention to his readers that while the difficulty with the vanishing of $\mathfrak{H}$ is avoided by using the tensor $\mathbb{H}^{\alpha\beta} = p^\alpha u^\beta - \eta^{\alpha\beta}\mathfrak{L}$, in order *"to obtain a Hamiltonian that has the meaning of energy, we must proceed differently"*.[55] The conclusion at this point is that there is no single covariant Hamiltonian that was a similar quantity to energy and does not vanish identically.

Note: Some readers might object that we can avoid the difficulties in finding a suitable covariant Hamiltonian by always working in the Lagrangian formalism. However, the Hamiltonian method is fundamental for both statistical mechanics and quantum theory.

Let us now look at another difficulty with the covariant formalism. The expression $\epsilon_0(\boldsymbol{E}^2 + c^2\boldsymbol{B}^2)/2$ that appears in the Hamiltonian (2.18) is replaced in the covariant formulation by the following tensor

$$T^{\alpha\beta} = \frac{1}{\mu_0}\left(F^{\alpha\lambda}F^{\beta}_{\lambda} - \frac{1}{4}\eta^{\alpha\beta}F_{\lambda\zeta}F^{\lambda\zeta}\right), \qquad (11.23)$$

so that $T^{00} = \epsilon_0(\boldsymbol{E}^2 + c^2\boldsymbol{B}^2)/2$. We saw in chapter 4 that the volume integral of T^{00} cannot be consistently interpreted as the energy of the electromagnetic field. The situation worsens in the covariant formulation because the rest of components of $T^{\alpha\beta}$ cannot be associated with properties of the electromagnetic field either and the quantity (11.23) cannot be accepted as the proper tensor for the electromagnetic field. There is a typical rule for transitioning from a noncovariant to a covariant formulation: if the noncovariant formulation has a problem, the covariant formulation generally has the same problem and more.

Note: The tensor $T^{\alpha\beta}$ is given different names in the literature: "*symmetric stress tensor*",[6] "*energy-momentum tensor*",[7] "*symmetric energy-momentum tensor*",[8] and "*stress-energy-momentum tensor (or stress-energy tensor, for short)*".[10] All the names are incorrect because (**i**) T^{00} is an energy density and not an energy, (**ii**) T^{0i} is c multiplied by the momentum density, and (**ii**) the true stresses are only described by the T^{ij} components when $i \neq j$.

Remember that in this chapter we only focus on the misconceptions that are unique to the covariant formulations, and that we have ignored all the misconceptions already discussed in the previous chapters. For example, the covariant Lagrangian $\mathfrak{L}$ and the associated pseudo-Hamiltonians are formally divergent and if we replace the four-potentials associated with the field by external ones $A^{\alpha} \rightarrow A^{\alpha,\text{ext}}$ to eliminate the infinities, then the resulting covariant equation of motion violates the conservation laws because the radiation reaction terms are missing from the equation, as we already saw in chapter 7. Similarly, if we add a covariant term $\int \mathrm{d}^4x(1/4\mu_0)F^{\alpha\lambda}F_{\alpha\beta}$ for the electromagnetic field to the $\mathfrak{L}$ defined in (11.21), we will end up with a covariant Lagrangian for a system of charges and field that has essentially the same difficulties that we discussed in chapter 4 for the noncovariant formulation of the same system.

As the main **conclusions** of this chapter we can offer the following list:

- The covariant formulation is redundant. For example, the electromagnetic field tensor has sixteen components but there are only six physical degrees of freedom. The standard notation is also ambiguous.

- Contrary to myth, the laws of nature do not need to be expressed in covariant form.
- There is no compact covariant formulation for the motion of a system of *N* charges.
- The so-called covariant Hamiltonians are not true Hamiltonians.
- As a typical rule, if the noncovariant formulation has a problem, the covariant formulation generally has the same problem and a few more.

—12

THE NOETHER THEOREM

Mathematician Noether proved several theorems during her lifetime, but what has become known among physicists and mathematicians as the Noether theorem is "*a general connection between symmetries and conserved quantities*"[54] for physical systems.

This theorem is sometimes poorly stated as telling "*us that conservation laws follow from the symmetry properties of nature*"[56] and is often praised as "*one of the most amazing and useful theorems in physics*",[56] and "*one of the most elegant, beautiful and powerful results in theoretical physics*",[57] because this "*theorem proves a deep relationship between symmetries and conserved quantities*".[57] Hanca, Tulejab, and Hancovac reproduce a quote from Weinberg in which he says that "*it is increasingly clear that the symmetry group of nature is the deepest thing that we understand about nature today*".[56]

We find in the cited literature an interesting collection of myths and exaggerations regarding symmetries and conservation laws, but before delving into the technical details, we would mention that physicists usually call Noether theorem to the first theorem in the paper "*Invariante Variationsprobleme*", published in 1918, in which she studies the group properties of Lagrangian actions $\mathfrak{I}$ and concludes that (translated from the german original):

> "*If the integral $\mathfrak{I}$ is invariant under a [finite continuous group] $\mathfrak{G}_\rho$ with ρ parameters, then there are ρ linearly independent combinations among the Lagrangian expressions which become divergences — and conversely, that implies the invariance of $\mathfrak{I}$ under a $\mathfrak{G}_\rho$. The theorem remains valid in the limiting case of an infinite number of parameters.*"

She also adds that "*the divergence equations often referred to of late as 'laws of conservation' are obtained*". The aforementioned paper was motivated by the problem of energy conservation in the general theory of relativity.

The Noether theorem can be expressed informally as:

> "*If a Lagrangian system has a continuous symmetry property under which the Lagrangian is invariant in form, then there are corresponding quantities whose values are constant in time.*"

After putting things in perspective with these briefs historical comments, we can begin our technical analysis of the theorem and the misconceptions found in the electrodynamics literature.

First, the Noether theorem does not provide a "*general connection between symmetries and conserved quantities*" as Charap claims.[54] The theorem has limited validity since it applies to continuous symmetries, but not to discrete symmetries. Being derived from Lagrangian actions, the theorem does not apply to systems that cannot be modeled with a Lagrangian alone, such as systems that require a Rayleigh dissipation function. Indeed, Wigner warned in 1954 against the naive identification of symmetry and conservation principles.

Secondly, concepts such as "*elegant*" and "*beautiful*", used by physicists like Pössel[57] to characterize the theorem, are subjective and would be removed from any discussion of physics. Thirdly, when Hanca, Tulejab, and Hancovac claim[56] that conservation laws follow from symmetries, they omit that the theorem proven by Noether offers a two-way correspondence between symmetries and *some* conservation laws. Indeed, we can start from *some* symmetries and derive the associated conservation laws or we can start from those laws and derive the symmetries. I have emphasized the word 'some' here, because we will see below that what Noether called the "*laws of conservation*" are not always true conservation laws.

Are symmetries the deepest thing that we understand about nature? Weinberg, as always, confuses the model with reality. There is no such thing like "*the symmetry group of nature*". Most of the symmetries that he mentions in the reference cited by Hanca, Tulejab, and Hancovac are properties of the standard model of particle physics, not properties of nature. Furthermore, Weinberg omits that energy is not conserved in the consensus model of cosmology, the Lambda-CDM model, just as he also omits that the time-symmetry of the standard model is only valid when we ignore dissipative processes and chaos.

Note: Weinberg mentions symmetries compatible with quantum mechanics and symmetries unique to the standard model of particle physics, which is firmly based on quantum field theory. Weinberg believes that quantum field theory is just quantum mechanics applied to fields, but this is not true, A detailed discussion of the differences between quantum field theory and quantum mechanics will be offered in another book in this series. Without going any further, the consensus model of cosmology does not conserve energy since this model is based on general relativity. Energy is not conserved in general relativity because this is a geometric theory where the energy of gravitation itself is not included in the equations.

Theoretical physicists such as Weinberg like symmetries because these simplify their work even when their work no longer resembles the real world. This reminds me of one of my favorite jokes, which goes like this:

> *"A physicist, an engineer, and a psychologist are hired as consultants for a dairy farm whose production has been below average. Each is given time to inspect the details of the operation before making a report.*
>
> *After a week, the engineer comes back with a report that says: 'If you want to increase milk production, you need bigger milk pumps and bigger tubes to suck out the milk'.*
>
> *The next to report is the psychologist, who proposes: 'You have to make the cows produce more milk. One way to do this is to make them calm and happy. Happy cows produce more milk. Paint the milking stalls green. This will make the cows think about pastures and happy fields. Additionally, more trees should be planted in fields to add diversity to the landscape for livestock while grazing and reduce boredom. They will be happy.'*
>
> *Finally, the physicist is called, who looked around the place and then drew a circle on the blackboard saying, 'Let me consider a spherical cow in a vacuum, emanating milk uniformly in all directions...'"*

The joke is that the physicist had the solution to the farmer's problem, but his solution only works for spherical cows in a vacuum. The lesson we must learn here is that a symmetry is a useful tool only for describing objects and processes that conform to that symmetry, but reality cannot be changed to conform to your favorite symmetry.

Griffiths[19] provides a table with four typical applications of the Noether theorem. I reproduce his table in 12.1. It is a fact that we do not need the Noether theorem to understand this table.

Symmetry	'Conservation' law
Translation in time	Energy
Translation in space	Momentum
Rotation	Angular momentum
Gauge transformation	Charge

Table 12.1: Symmetries and 'conservation' laws. As shown in the text, the conservation of a quantity is often confused with its constancy.

Consider the first entry in the table. We know, from Hamiltonian mechanics, that

$$\frac{\mathrm{d}H}{\mathrm{d}t} = \left(\frac{\partial H}{\partial t}\right)_{\boldsymbol{p},\boldsymbol{r}}, \tag{12.1}$$

which implies that the Hamiltonian H is a constant of motion, $(\mathrm{d}H/\mathrm{d}t) = 0$, when it does not depend *explicitly* on time or, what is the same, the Hamiltonian does not change if we shift time by an arbitrary amount Δt, since it does not depend on time. The opposite is also true (if the Hamiltonian is invariant to translations in time then it is a constant of motion), and we conclude that

$$H = \text{const} \quad \longleftrightarrow \quad H(t) = H(t + \Delta t), \tag{12.2}$$

where the double arrow means that we can assume the condition on the left and derive the one on the right and vice versa. Thus, we have derived the first entry in table 12.1 without using the Noether theorem. Let us derive the rest of the entries.

Consider now a system for which the momentum $\boldsymbol{p}$ is a constant of motion; that is, $(\mathrm{d}\boldsymbol{p}/\mathrm{d}t) = 0$. If we use the second of the Hamilton equations

$$\frac{\mathrm{d}\boldsymbol{p}}{\mathrm{d}t} = -\left(\frac{\partial H}{\partial \boldsymbol{r}}\right)_{\boldsymbol{p}}, \tag{12.3}$$

we obtain that the partial derivative is zero and this implies that the Hamiltonian of this system is invariant to arbitrary changes $\Delta \boldsymbol{r}$ in position. Conversely, the momentum is constant in time if the Hamiltonian is invariant to the aforementioned change. In formal terms,

$$\boldsymbol{p} = \text{const} \quad \longleftrightarrow \quad H(\boldsymbol{r}) = H(\boldsymbol{r} + \Delta \boldsymbol{r}). \tag{12.4}$$

Analogously, we can demonstrate that the angular momentum $\boldsymbol{L}$ is constant when the Hamiltonian is invariant to the application of a rotation matrix $\boldsymbol{\Lambda}$ to the coordinates

$$\boldsymbol{L} = \text{const} \quad \longleftrightarrow \quad H(\boldsymbol{r}) = H(\boldsymbol{\Lambda r}). \tag{12.5}$$

That is, we have derived the first three entries in the table provided by Griffiths without mentioning Noether and her theorem. The last entry can be derived in a similar way, because gauge transformations can be seen as a special case of the invariance of the Hamiltonian equations of motion under a canonical transformation. In these cases, the theorem does not provide any new result that had not been known eighty-five years before.

The Noether theorem is a very interesting result but its usefulness and physical meaning are exaggerated in most of the modern physics literature. Moreover, in the initial part of this chapter I wrote that what Noether called the *"laws of conservation"* are not always conservation laws. Now we will see why.

But first of all, a small conceptual digression. An extensive property is a physical quantity whose value depends on the amount of substance, while an intensive property does not depend on it. Charge, heat capacity, mass, volume, and momentum are typical examples of extensive quantities, while charge density, temperature, mass density, specific volume, and velocity are of intensive quantities.

Consider an extensive quantity G whose density is denoted by g; that is, $G = \int g \, \mathrm{d}V$. The change in the amount of G in a volume V is the sum of the net flow of G into the volume and the production of G within that volume.

If $\boldsymbol{J}_G$ is the current density (the amount of G that leaves the volume through unit area per unit time), then the change in G due to the flow is $\int \boldsymbol{J}_G \cdot \mathrm{d}\boldsymbol{S}$ with $\mathrm{d}\boldsymbol{S}$ the vector representing the infinitesimal area element, as illustrated in figure 12.2. If $\mathbb{P}_G$ is the amount of G produced per unit volume per unit time, then the change in G due to production is $\int \mathbb{P}_G \, \mathrm{d}V$. The total change in G in the volume can be written as

$$\frac{\mathrm{d}}{\mathrm{d}t}\int g \, \mathrm{d}V = \int \mathbb{P}_G \, \mathrm{d}V - \int \boldsymbol{J}_G \cdot \mathrm{d}\boldsymbol{S}. \tag{12.6}$$

This is sometimes called a balance equation and it is represented graphically in the figure below.

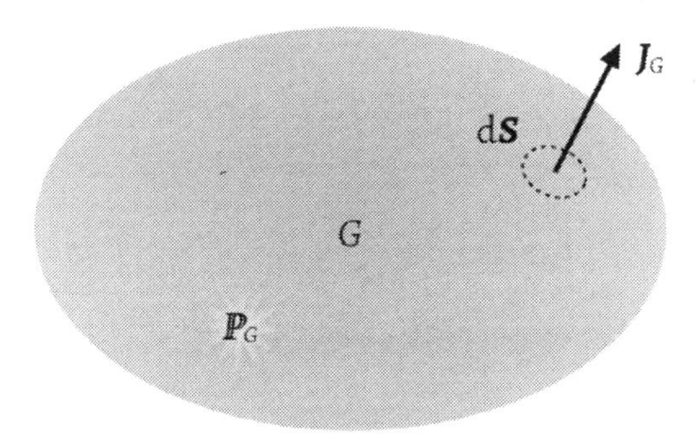

Figure 12.2: The change in the amount of G in a volume V is due to the net flow $\int \boldsymbol{J}_G \cdot \mathrm{d}\boldsymbol{S}$ entering or leaving the volume across its boundaries plus, $\int \mathbb{P}_G \, \mathrm{d}V$, the amount of G produced in the volume.

We say that a quantity is *conserved* if its production in the volume is zero, $\int \mathbb{P}_G \, \mathrm{d}V = 0$, because the change in the amount of the quantity is due solely to the flows through the surface, since the quantity cannot be created or destroyed in the volume.

Equation (12.6) is expressed in integral form. If we introduce the time derivative into the integral

$$\frac{\mathrm{d}}{\mathrm{d}t}\int g\,\mathrm{d}V = \int \frac{\partial g}{\partial t}\,\mathrm{d}V, \tag{12.7}$$

and use the divergence theorem

$$\int \boldsymbol{J}_G \cdot \mathrm{d}\boldsymbol{S} = \int \nabla \cdot \boldsymbol{J}_G\,\mathrm{d}V, \tag{12.8}$$

we can derive a local formulation of the balance,

$$\frac{\partial g}{\partial t} = \mathbb{P}_G - \nabla \cdot \boldsymbol{J}_G. \tag{12.9}$$

Equation (12.6) applies to the entire volume while (12.9) is valid for any point within the volume. Furthermore, if we define a four-current density $J^\alpha_G = (cg, \boldsymbol{J}_G)$, we can rewrite (12.9) in an explicitly covariant form

$$\partial_\alpha J^\alpha_G = \mathbb{P}_G. \tag{12.10}$$

We stated above that a quantity G is conserved when its production in the volume is zero. This is equivalent to setting $\mathbb{P}_G = 0$, because if the quantity is conserved at any point, then it is conserved in the entire volume since $\int \mathbb{P}_G\,\mathrm{d}V = 0$.

We have derived the balance equations and the conservation condition for a scalar quantity. If instead we had considered vector or higher rank tensor quantities, we had derived higher rank expressions. For example, the explicitly covariant form of the local balance law for the 'stress-energy-momentum' tensor is

$$\partial_\alpha T^{\alpha\beta} = \mathbb{P}^\beta. \tag{12.11}$$

With this brief description of balance equations available in our hands, we can return to the Noether theorem and analyze whether it really provides a connection between symmetries and conserved quantities.

The symmetries associated to this theorem are usually classified into internal and spacetime symmetries, with the internal symmetries do not involving manipulations of spatial and temporal coordinates. Charap offers as an illustration of the application of the theorem to electrodynamics: the connection between gauge symmetry and charge conservation.[54] This is an example of an internal symmetry.

Charap assumes the following action for the electromagnetic field

$$\mathfrak{I} = \int L\,\mathrm{d}t = \iint \left(-\frac{1}{4\mu_0} F_{\alpha\beta} F^{\alpha\beta} - J^{\alpha} A_{\alpha} \right) \mathrm{d}V \mathrm{d}t. \tag{12.12}$$

The term in parentheses is a Lagrangian density and if we integrate this density over the volume we obtain a Lagrangian for the electromagnetic field that corresponds to the interpretation (4.3). Let us ignore that this action cannot be consistently associated to the field (see the discussion in chapter 4) and consider the variation of this action under a transformation $A_\alpha \to A_\alpha + \partial_\alpha \chi$.

Ignoring the term proportional to $F_{\alpha\beta}F^{\alpha\beta}$, the action varies as

$$\Delta \mathfrak{I} = -\iint J^{\alpha} \Delta A_{\alpha}\, \mathrm{d}V \mathrm{d}t, \tag{12.13}$$

in which $\Delta A_\alpha = \partial_\alpha \chi$. Assuming that "*the sources are confined to a finite region*",[54] the integral can be reduced to

$$\Delta \mathfrak{I} = -\iint (\partial_\alpha J^{\alpha}) \chi\, \mathrm{d}V \mathrm{d}t. \tag{12.14}$$

By setting the variation of the integral to zero, $\Delta \mathfrak{I} = 0$, we obtain

$$\partial_\alpha J^{\alpha} = 0, \tag{12.15}$$

which is precisely (11.9). This is a true conservation law, because it corresponds to the local balance equation (12.10) when there is no charge production.

We can see that the law of conservation of charge has been derived, Noether-style, after a series of assumptions. Statements like "*this theorem tells us that conservation laws follow from the symmetry properties of nature*" are a complete exaggeration, because to derive the law of conservation of charge, we had to assume that the physical system of interest is described by an action, that the action does not vary under the symmetry, and that the system satisfies certain boundary conditions. We had to assume three conditions, when it would be much easier to assume the conservation law (11.9) directly.

There is a hidden assumption in the derivation given by Charap in his textbook,[54] and that is that $F_{\alpha\beta}F^{\alpha\beta}$ does not vary under the gauge transformation. If we use the same boundary conditions used by Charap, we can show that $F_{\alpha\beta}F^{\alpha\beta} = -2\mu_0 J^\alpha A_\alpha$ and instead of (12.14), we obtain

$$\Delta\mathfrak{I} = -\frac{1}{2}\iint (\partial_\alpha J^\alpha)\chi \, \mathrm{d}V\mathrm{d}t. \tag{12.16}$$

This correction does not change the final result, because we can still obtain the same conservation law since $(1/2)\cdot 0 = 0$, but it illustrates once again the inconsistencies that appear when we try to identify the dynamic properties of the electromagnetic field in the conventional theory.

Note: The above correction is irrelevant for the purposes of obtaining the charge conservation law from the Noether theorem, but it is crucial to obtain the correct equations of motion for a system of N interacting charges.

After this discussion of gauge symmetry and charge conservation, let us consider what the Noether theorem has to tell us about the rest of entries in the table 12.1. Translations in time, translations in space, and rotations are examples of spacetime symmetries.

We will consider only spacetime translations, because rotations can be studied in a similar way. Translation invariance of the electromagnetic field gives the following law

$$\partial_\alpha T^{\alpha\beta} = -F^{\alpha\beta} J_\alpha, \tag{12.17}$$

with $T^{\alpha\beta}$ the 'stress-energy-momentum' tensor (11.23). The above law, derived from the Noether theorem, is not a conservation law because if we compare it with the generic form (12.11), we find a non-zero production term $\mathbb{P}^\beta = -F^{\alpha\beta} J_\alpha$.

Only for a free electromagnetic field do we obtain a true conservation law, since in absence of charges $J^\alpha = 0$ and (12.17) reduces to $\partial_\alpha T^{\alpha\beta} = 0$. With this example, we have just confirmed that the Noether theorem does not always associate a conservation law with a symmetry, as Wigner anticipated.

Note: Wigner studied a simple equation of motion for a mechanical system without Lagrangian description but with time translation and spatial isotropy symmetries, and discovered that there are no conservation laws, in Noether's sense, for energy and angular momentum. We have arrived at essentially the same conclusion, but for a Lagrangian system.

A very common misconception found in the literature on electromagnetism is that a quantity is conserved only when it is constant, "*the existence of such a constant implies a conservation law, which we then identify*",[56] and if we review all the examples worked by Possel[57] we can see that by energy conservation he means that E is a constant. This misconception is also found in mechanics textbooks:[37]

> "*Functions of the dynamical variables and their time derivatives that remain constant during the motion are called conserved quantities or constants of the motion. Noether's theorem reveals how the symmetries of the Lagrangian can be used to construct constants of the motion from the Lagrangian.*"

But before looking at those textbooks, let me present an alternative formalism for the balance equations. We will multiply both sides of (12.6) by $\mathrm{d}t$ and using the notation $\mathrm{d}_i G = \int \mathbb{P}_G \,\mathrm{d}V\mathrm{d}t$ and $\mathrm{d}_e G = -\int \boldsymbol{J}_G \cdot \mathrm{d}\boldsymbol{S}\,\mathrm{d}t$, along with $G = \int g\,\mathrm{d}V$, we can rewrite it as

$$\mathrm{d}G = \mathrm{d}_i G + \mathrm{d}_e G, \tag{12.18}$$

with $\mathrm{d}_i G$ and $\mathrm{d}_e G$ representing the production and flow of the quantity, respectively. We call this the *de Donder notation* after Théophile Ernest de Donder. In this new notation, a quantity G is conserved when $\mathrm{d}_i G = 0$.

We can analyze the mechanics textbooks now. For simplicity, consider a simple system with a single degree of freedom and arbitrary Lagrangian $L = L(\boldsymbol{r}, \boldsymbol{v})$. If we apply the Noether theorem to time translations, we obtain that the following quantity,

$$E = \left(\frac{\partial L}{\partial \boldsymbol{v}}\right)_{\boldsymbol{r}} \boldsymbol{v} - L, \tag{12.19}$$

is a constant of motion. This constant is the energy. However, the law of the conservation of energy is not $\mathrm{d}E = 0$. As we saw previously, a quantity is conserved when its production is zero. Using the de Donder notation, the integral form of the balance law for energy is

$$\mathrm{d}E = \mathrm{d}_i E + \mathrm{d}_e E, \tag{12.20}$$

and the law of conservation of energy is $\mathrm{d}_i E = 0$. Introducing this conservation condition into the above balance law, we obtain

$$\mathrm{d}E = \mathrm{d}_e E. \tag{12.21}$$

This reduced form of the balance law states that the energy of any system is still conserved when it is not a constant, $\mathrm{d}E \neq 0$. Indeed, the law of conservation of the energy states that this quantity cannot be created or destroyed, but it does not prevent energy from flowing between two or more systems.

In thermodynamics, the flow term $\mathrm{d}_e E$ is usually given by heat and work for closed systems, and this is why the first law of thermodynamics, the law of conservation of energy, is $\mathrm{d}E = \mathrm{d}Q + \mathrm{d}W$.

Note: For open systems, there is an additional contribution due to the flow of matter and the law is $\mathrm{d}E = \mathrm{d}Q + \mathrm{d}W + \mathrm{d}R$. Furthermore, we are using exact differentials because we are using the formalism of modern thermodynamics. Classical thermodynamics only deals with equilibrium states and, as a consequence, must rely on the so-called inexact or imperfect differentials; that is, $\mathrm{d}E = đQ + đW + đR$.

Energy is always conserved since $\mathrm{d}_i E = 0$ holds for both isolated and nonisolated systems. However, energy is a constant of motion only for isolated systems because for them $\mathrm{d}_e E = 0$.

Thus, the condition (12.2) associated to the Noether theorem implies that energy is constant for systems that satisfy time-translation symmetry, but this condition is not a conservation law. It is not strange then that energy is also conserved when time symmetry is not satisfied. This is just what happens in those dissipative systems that evolve under the constraint $\mathrm{d}E \leq 0$. The energy of the system is not invariant under time translations, $E(t) \neq E(t + \Delta t)$, but energy is conserved because the change in the energy of the system is due exclusively to a flow of energy with the surroundings; that is, energy is not created or destroyed.

Note that the Noetherian literature that identifies a conserved quantity only with a quantity that is a constant of the motion is not only ignoring the history of physics since the 19th century, but it is also applying an inconsistent criterion to the identification of conservations laws. For the Noetherian literature, energy is conserved only when it is constant, but charge is a conserved quantity –see (12.15)– because its production is zero, $\mathbb{P}_Q = 0$, or in de Donder notation when $\mathrm{d}_i Q = 0$.

We are faced here with a serious inconsistency of the Noetherian literature. How could energy E and momentum $\boldsymbol{p}$ be conserved *only* when they are constants of the motion, but charge Q could be conserved even when it is not constant? Energy is conserved only when $\mathrm{d}E = 0$, but charge is conserved even when $\mathrm{d}Q \neq 0$? This does not make any sense and reflects a fundamental deficiency of the theorem and how much of modern physics is being developed by ignoring the depth nature of physical phenomena and replacing it with superficial geometric and symmetric approaches.

The main **conclusions** of this chapter are the following:

- What physicists often call the Noether theorem is the first theorem of her paper *Invariante Variationsprobleme*.
- The Noether theorem does not provide a "*general connection between symmetries and conserved quantities*".
- There is no such thing like "*the symmetry group of nature*". Gauge symmetries and similar symmetries are properties of specific models of nature, not a property of nature itself.
- We do not need the Noether theorem to understand when properties such as energy and momentum are constants of motion.
- The Noetherian literature confuses a conserved quantity with a quantity that is a constant of the motion.
- A quantity is conserved only when its production is zero, not when the quantity is a constant. The laws of conservation of energy, momentum, and charge are $\mathbb{P}_E = 0$, $\mathbb{P}_{\boldsymbol{p}} = 0$, and $\mathbb{P}_Q = 0$, respectively or $\mathrm{d}_iE = 0$, $\mathrm{d}_i\boldsymbol{p} = 0$, and $\mathrm{d}_iQ = 0$ in the de Donder notation.

—13

MISCELLANY

So far, we have presented a collection of misconceptions that are unique to the formulations of classical electrodynamics found in leading textbooks on the subject. This chapter is dedicated to exploring the frontiers between electrodynamics and other disciplines of physics such as general relativity, thermodynamics, and quantum theory, and the new misconceptions that appear at the frontiers.

GENERAL RELATIVITY

In this section we use, for reasons of familiarity, the covariant formulation presented in chapter 11. We also use the Einstein notation.

Special relativity is based on a flat spacetime. General relativity considers a spacetime that can be curved and the line element (11.6) is replaced by

$$ds^2 = dx^\alpha g_{\alpha\beta} dx^\beta. \tag{13.1}$$

Note: Just as the standard terminology used in the electromagnetic literature is misleading when referring to "*electromagnetic fields*", because there is a single electromagnetic field that extends throughout all of space. The literature on general relativity suffers from a similar deficiency when the authors refer to "*gravitational fields*" and "*spacetimes*".[58–61] This is because general relativity assumes that there is a single spacetime that extends throughout space and, contrary to a widespread myth, there is no gravitational field in general relativity, since this is a metric theory.

Most covariant equations in flat spacetime can be generalized directly to a curved spacetime by use of the "*minimal substitution rule*"[61] in which $\eta_{\alpha\beta} \rightarrow g_{\alpha\beta}$ and $\partial_\beta \rightarrow \nabla_\beta$, with ∇_β the partial derivative operator associated with $g_{\alpha\beta}$.

Note: The notation used here is standard but a bit misleading. The covariant partial derivative ∇_β would not be confused with the components ∇_i of the three-dimensional gradient operator $\mathbf{\nabla}$ in flat spacetime.

With the help of the minimal substitution rule, the Maxwell equations in flat spacetime (11.10) and (11.13) generalize to[60,61]

$$\nabla_\beta F^{\alpha\beta} = \mu_0 J^\alpha \tag{13.2}$$

and

$$\nabla_\zeta F^{\alpha\beta} + \nabla_\alpha F^{\beta\zeta} + \nabla_\beta F^{\zeta\alpha} = 0. \tag{13.3}$$

Similarly, the Minkowski-Lorentz equation of motion (11.8) becomes, in general relativity,[60]

$$\frac{\mathrm{d}(mu^\alpha)}{\mathrm{d}\tau} + m\Gamma^\alpha_{\zeta\lambda} u^\zeta u^\lambda = qF^{\alpha\beta} u_\beta, \tag{13.4}$$

where $\Gamma^\alpha_{\zeta\lambda}$ are "*the Christoffel symbols*".[58–60]

Some equations cannot be obtained from the minimal substitution rule, but require a more detailed analysis of the consequences of introducing a curved spacetime in the description of electromagnetic phenomena. An example is the Maxwell equations in the potential formulation (11.14), whose generalization to curved spacetime in the Lorenz gauge $\nabla_\alpha A^\alpha = 0$ is[60]

$$\nabla_\beta \nabla^\beta A^\alpha - R^\alpha_\beta A^\beta = -\mu_0 J^\alpha, \tag{13.5}$$

with R^α_β *"the Ricci tensor"*.[60,61] If we had only performed the substitution $\partial_\beta \to \nabla_\beta$ in (11.14), we would miss the curvature term $R^\alpha_\beta A^\beta$ in (13.5). In this case, we can decide in favor of equation (13.5) over the alternative equation without the curvature term because (13.5) implies charge conservation in curved spacetime, while the alternative equation conflicts with it. The reason why the correction to the naive application of the minimal substitution rule involves a curvature term is that the derivatives do not commute in a curved spacetime, $\nabla_\alpha \nabla_\beta \neq \nabla_\beta \nabla_\alpha$, while they do in flat spacetime, $\partial_\alpha \partial_\beta = \partial_\beta \partial_\alpha$.

All the misconceptions of the ordinary formulations of electrodynamics in flat spacetime remain when electrodynamics is combined with general relativity. I am not going to repeat everything said in the previous chapters about divergences, violation of the laws of conservation and inertia, the nonrelativistic limit, many-body motion, interpretation issues, and more. In the remainder of this section I will only discuss the misconceptions associated exclusively with the use of a curved spacetime in electrodynamics.

One of the problems of general relativity is the ambiguity regarding the definition and characterization of reference frames. In special relativity, spacetime is both constant and common to the entire universe, but general relativity introduces a four-dimensional dynamical space $x^\zeta = (x^0, x^1, x^2, x^3)$ whose metric tensor $g_{\alpha\beta}(x^\zeta)$ can vary in space and time. This has an important consequence for measuring distances to distant objects (for example to stars and distant planets), because while the pseudotemporal coordinate (let us call it x^0) can be measured directly with the observer's clock, the pseudospatial coordinates cannot be measured, but they can only be calculated *after* knowing the metric. However, the value of the metric at a given point depends on the distribution of matter and radiation and, thus, it depends on distances. This is a chicken and egg type problem.

In practical applications, physicists and astronomers assume a given metric for distant localizations, then use this metric to calculate the distances to objects at that localization (for example, using light signals) and finally calculate the metric from the distribution of matter and radiation. If the obtained metric coincides with the initially assumed metric, then the procedure stops here; otherwise, it is necessary to assume a new metric and repeat the procedure. Note that this procedure only provides a consistency check (that the metric assumed to interpret the observations is compatible with the metric calculated from the observations), but this procedure does not guarantee that we have obtained the correct metric, since two or more metrics could be compatible with the same observations. The bottom line is that electrodynamics equations such as (13.2), (13.3), (13.4) and (13.5) are based on curved geometries that are generally not known, but only assumed. Electrodynamics in curved spacetime does not provide an unambiguous combination of electromagnetism and gravitation.

Let us now see another difficulty. In flat spacetime, it is not possible to write a consistent and complete expression for the energy of a system of interacting charges (see chapter 8). In general relativity, not even the energy of a single charge is well defined:[61]

> *"Because spacetime is curved, there is no well defined notion of vectors at different points being parallel; parallel transport is curve dependent. Thus, there is no natural 'global family' of inertial observers, and a given observer cannot, in general, define the energy of a distant particle."*

As a consequence, it is not possible to write a complete and consistent expression for the energy of a system of charges interacting both electromagnetically and gravitationally. It is possible to generalize the Lagrangian (2.17) and the Hamiltonian (2.18) to curved spacetimes but the gravitational energy of the system of charges and electromagnetic field cannot be defined and the combined description of electromagnetism and gravitation is clearly incomplete.

The tensor (11.23) can be generalized to curved spacetimes as

$$T^{\alpha\beta} = \frac{1}{\mu_0}\left(F^{\alpha\lambda}F^{\beta}_{\lambda} - \frac{1}{4}g^{\alpha\beta}F_{\lambda\zeta}F^{\lambda\zeta}\right). \qquad (13.6)$$

This generalization has the same difficulties of interpretation as (11.23). Not only can the new tensor (13.6) not be attributed to an electromagnetic field, but it now incorporates the difficulties inherent in using a curved spacetime. Moreover, it is not possible to combine (13.6) with a similar tensor $T^{\alpha\beta}$ for the gravitational 'field' to obtain a unification of gravitation and electromagnetism. I put the above term in single quotes because, strictly speaking, there is no gravitational field in general relativity. General relativity is a metric theory in which gravitation is associated with curved spacetime, and spacetime is a manifold, not a field. Thus, contrary to what Straumann claims, the pseudo-Hamiltonian

$$\mathcal{H} = \frac{1}{2}(p^{\alpha} - qA^{\alpha})\, g_{\alpha\beta}\, (p^{\beta} - qA^{\beta}) \qquad (13.7)$$

does not describe *"the motion of a charged particle with charge [q] in a gravitational field"*.[60] Even if we expand the metric tensor $g_{\alpha\beta}$ around a Minkoswki background, $g_{\alpha\beta} = \eta_{\alpha\beta} + h_{\alpha\beta}$, the deviation $h_{\alpha\beta}$ from flatness does not describe any gravitational field. In fact, general relativity states that the quantity $h_{\alpha\beta}$ is not observable, since it represents a deviation from an unobservable flat background.

Note: Any attempt to find an expression similar to (13.6) but for the gravitational 'field' gives absurd results (pseudotensors, nonconservation,...) because the geometric approach used in general relativity omits crucial physical aspects of the gravitational interaction. The technical details of why general relativity cannot be considered the theory of a gravitational field are rather complex and beyond the scope of this section. The details are included in another volume of this series on *common misconceptions in physics*: the volume dedicated to relativity.

THERMODYNAMICS

According to Haase, "*thermodynamics, in its most general form, is the theory of those aspects of the macroscopic behavior of matter not covered by mechanics and electrodynamics*".[62] This means that the electrodynamics literature generally ignores thermodynamic corrections to electromagnetic laws.

The equations that we find in electrodynamics textbooks use pure electromagnetic quantities such as $\boldsymbol{E}$ and $\boldsymbol{B}$, charge and current densities, and some mechanical quantities such as velocity and energy, but these equations ignore thermodynamic quantities such as heat, entropy, and temperature. We can develop extensions of the usual electromagnetic laws if we do not ignore them.

For most substances, the current density $\boldsymbol{J}$ is proportional to the force per unit charge, $\boldsymbol{J} = \sigma(\boldsymbol{F}/q)$, where σ is the conductivity of the material medium. Using the Lorentz force law (2.16), we obtain

$$\boldsymbol{J} = \sigma(\boldsymbol{E} + \boldsymbol{v} \times \boldsymbol{B}). \tag{13.8}$$

When the velocity of the charges is small enough, the second term in the equation can be ignored and

$$\boldsymbol{J} = \sigma \boldsymbol{E}. \tag{13.9}$$

This equation is called the Ohm law[5,6,10] or the generalized Ohm law.[8] A basic thermodynamic generalization of this electrodynamic law is

$$\boldsymbol{J} = \omega \left(\frac{\boldsymbol{E}}{T} \right) + \lambda \nabla \left(\frac{1}{T} \right), \tag{13.10}$$

where ω and λ are two material coefficients, T is the temperature of the medium, and $\omega = \sigma T$. We can see that while the Ohm law (13.9) predicts zero currents when there is no electromagnetic field, the thermoelectric law (13.10) states that the current will not be zero if a temperature gradient exists. In fact, the Seebeck effect consists on generating an electromotive force by means of two dissimilar metal wires at different temperatures.

Note: Of course, there is no problem in teaching simplified models of nature to students, especially in the early stages of their education. The problem arises when teachers and textbooks present approximate laws as if they were valid in any empirical situation.

The class of thermoelectric phenomena described by laws such as (13.10) is ignored in most electrodynamics texts,[6,8,10] and only Griffiths barely mentions that, in typical electric circuits, the quantity $\boldsymbol{E}$ in (13.9) must be replaced by $(\boldsymbol{E} + \boldsymbol{f})$, where the physical agency responsible for the additional force per charge $\boldsymbol{f}$ can be anyone of many different things: "*in a battery it's a chemical force; in a piezoelectric crystal mechanical pressure is converted into an electrical impulse; in a thermocouple it's a temperature gradient that does the job; in a photoelectric cell it's light; and in a Van de Graaff generator the electrons are literally loaded onto a conveyor belt and swept along*".[5]

Using the identity $\nabla(1/T) = -(\nabla T)/T^2$ on the second term of (13.10) and using the relationship between ω and σ, we obtain

$$\boldsymbol{J} = \sigma \left(\boldsymbol{E} - \frac{\lambda}{\omega T} \nabla T \right), \tag{13.11}$$

and we can identify the value of the additional force per charge with a gradient of temperature; that is, $\boldsymbol{f} = -(\lambda/\omega T)\nabla T$. The prefactor $(\lambda/\omega T)$ is the Seebeck coefficient, also known as thermopower or thermoelectric sensitivity. We can see through thermodynamic reasoning that this coefficient depends on temperature because the physical agent for the appearance of this additional force per charge is the existence of an inverse temperature gradient $\nabla(1/T)$ as shown by (13.10).

Expression (13.10) can be generalized further, since chemical processes such as diffusion and reactions also produce contributions to the electromagnetic current density $\boldsymbol{J}$. In this section we have only scratched the surface of the thermodynamics corrections to electrodynamics. Contrary to what is stated in many electrodynamics textbooks, the Maxwell equations combined with the Lorentz force equation do not form *"the basis of all classical electromagnetic phenomena"*.[6]

So far we have considered misconceptions that are the result of ignoring what thermodynamics has to say about electromagnetic phenomena, but we can also encounter misconceptions that are the result of taking the thermodynamics literature too seriously. I will explain myself better. The thermodynamics literature also contains misconceptions that often infiltrate other disciplines as well.

A complete discussion of the misconceptions in thermodynamics is beyond the scope of this section (another book in this series will be dedicated to thermodynamics), but I will give an example drawn from the literature on the so-called thermodynamics of irreversible processes whose acronym is TIP. For an isothermal volume element in which there is no diffusion or chemical reactions, and gravitational effects are negligible, the quantity

($\boldsymbol{J}\cdot\boldsymbol{E}/T$) is interpreted in the standard literature as the entropy production in the volume element due to electrical phenomena. However, this quantity cannot be a production since it is obtained after assuming that the volume is in internal equilibrium and by definition the production of entropy is zero at equilibrium.

QUANTUM THEORY

Jackson states in the introductory chapter of his textbook:[6]

> *"From the point of view of the standard model, classical electrodynamics is a limit of quantum electrodynamics (for small momentum and energy transfers, and large average numbers of virtual or real photons)."*

This is a myth, because we cannot derive one from the other. A detailed study of classicality and the correspondence principle is beyond the scope of this small section, but I will quote what Rohrlich has said on this topic:[63]

> *"Most physicists probably subscribe to the notion that classical (nonquantum) electrodynamics (CED) can be derived from QED by a suitable limiting process. This notion is logically very reasonable and in my view philosophically necessary because QED without CED is incomplete. This is easily seen by analyzing a measurement process in QED. And yet, there does not exist to date a clean proof of this limit."*

Rohrlich also provides us with a list of unresolved issues:

> *"What becomes of the pair production effects (for example, virtual creation and annihilation) in the classical limit? What becomes of radiative corrections in the classical limit? What becomes of the soft photon emission that necessarily accompanies all interactions of a charged particle not in a bound state?"*

There are other crucial questions that Rohrlich does not ask, such as why velocities in quantum electrodynamics are given in terms of Dirac alpha operators, $\hat{\boldsymbol{v}} = c\hat{\boldsymbol{\alpha}}$, when these operators are unphysical and predict absurd phenomena, including that electrons always move at the speed of light?

Indeed, using the relationship between the momentum of a charge and its velocity, the Lorentz equation (6.11) can be written as

$$\frac{\mathrm{d}\boldsymbol{p}}{\mathrm{d}t} = q\left(\boldsymbol{E} + \boldsymbol{v} \times \boldsymbol{B}\right) + q\frac{\mathrm{d}\boldsymbol{A}}{\mathrm{d}t}, \tag{13.12}$$

and the quantum electrodynamics version is reported to be

$$\frac{\mathrm{d}\hat{\boldsymbol{p}}}{\mathrm{d}t} = q\left(\hat{\boldsymbol{E}} + c\hat{\boldsymbol{\alpha}} \times \hat{\boldsymbol{B}}\right) + q\frac{\mathrm{d}\hat{\boldsymbol{A}}}{\mathrm{d}t}. \tag{13.13}$$

where $q\left(\hat{\boldsymbol{E}} + c\hat{\boldsymbol{\alpha}} \times \hat{\boldsymbol{B}}\right)$ is considered the quantum analogue of the Lorentz force. However, Feynman warns us that "*since there is no direct connection between this equation and* **v**, *it does not lead directly to Newton's equations in the limit of small velocities and hence is not completely acceptable as a suitable analogue*".[64]

Another myth found in Jackson's textbook is his claim that "*the standard model gives a coherent quantum-mechanical description of electromagnetic, weak, and strong interactions*".[6] This is wrong because the standard model is based on quantum field theory, which is incompatible with quantum mechanics, as Dirac denounced on more than one occasion. We can find a similar misconception in Griffiths' text: "*relativistic quantum mechanics is known as quantum field theory*".[5] No, the so-called relativistic quantum mechanics and quantum field theory are two disjoint theories and the more rigorous quantum field theory books even mention some of the fundamental differences with quantum mechanics such as that position is an observable in quantum mechanics but it is not observable in quantum field theory.[65]

Jackson also states that "*quantum electrodynamics, in turn, is a consequence of a spontaneously broken symmetry in a theory in which initially the weak and electromagnetic interactions are unified and the force carriers of both are massless*".[6] However, this is another myth because there is no real unification in the so-called electroweak theory, since this theory introduces two force carriers instead of only one as would correspond to a true unification.

Many modern physicists spread the unification myth among students and the general public. Feynman was one of the few who openly opposed the myth of the unification of interactions and always referred to the standard model as a collection of different theories. Crease, a philosopher and historian of science, once visited the *California Institute of Technology* (*Caltech*) to interview Feynman, and this is the discussion the two had about unification and the standard model:[66]

> **Crease>** "*Your career spans the period of the construction of the standard model*".
> **Feynman>** "*The standard model*" (he repeated dubiously).
> **Crease>** "SU(3) × SU(2) × U(1). *From renormalization to quantum electrodynamics to now?*".
> **Feynman>** "*The standard model, standard model*" (he said) "*The standard model—is that the one that says that we have electrodynamics, we have weak interaction, and we have strong interaction? Okay. Yes*".
> **Crease>** "*That was quite an achievement, putting them together*".
> **Feynman>** "*They're not put together*".
> **Crease>** "*Linked together in a single theoretical package?*".
> **Feynman>** "*No*".
> **Crease>** "*What do you call* SU(3) × SU(2) × U(1)*?*".
> **Feynman>** "*Three theories*" (he said) "*Strong interactions, weak interactions, and electromagnetic [...] The theories are linked because they seem to have similar characteristics*".

Jackson is not the only physicist that reproduces the myth of the electroweak unification, Wald writes about how *"the electromagnetic field together with the W and Z fields comprise a unified 'electroweak gauge field' that describes both the electromagnetic and weak interactions."*[10] Griffiths includes in his own textbook[5] a section titled *"The Unification of Physical Theories"* in which he not only repeats this myth but also promotes the scam of superstring theory as a candidate for the unification of all interactions.

Note: String theory was born as a genuine research project to understand hadrons and the strong interaction; this project failed, but instead of being abandoned the theory was modified and modified again until it finally became the biggest scam in the history of physics. In fact, the modern version of string theory is no longer about strings, but the name is kept for marketing purposes.

We have seen in previous chapters that classical electrodynamics is usually formulated as a field theory, but there is also an action-at-a-distance formulation that removes the electromagnetic field. However, Wald wants to convince us that fields are fundamental at the quantum level:[10]

> *"Thus, although there are circumstances where one could take the view that electromagnetic fields are produced by charges, it is far healthier to think of the electromagnetic field and charged matter as independent entities that interact via Maxwell's equations and eqs. [...]. Indeed, the view that electromagnetic fields are produced by charges is particularly untenable in quantum field theory, since it is essential for the understanding of such phenomena as the vacuum fluctuations of the electromagnetic field that the electromagnetic field has its own dynamical degrees of freedom, independently of the existence of charged matter."*

The above statement is doubly wrong. It is incorrect because there is an action-at-a-distance formulation of quantum electrodynamics (a quantum version of the classical theory of Wheeler and

Feynman) that eliminates the electromagnetic field and uses only charged particles.[67] Wald is also wrong because Casimir forces, which are sometimes taken as evidence for the existence of fluctuations in the vacuum associated to quantum field theory, can not only be calculated without any reference to vacuum fluctuations,[68] but it has recently been shown that the Casimir forces cannot originate from the vacuum state of the electromagnetic field.[69]

But there are more errors in literature. In a section of his introductory book on elementary particles, Griffiths writes:[19]

> *"Quantum electrodynamics (QED) is the oldest, the simplest, and the most successful of the dynamical theories; the others are self-consciously modeled on it. So I'll begin with a description of QED. All electromagnetic phenomena are ultimately reducible to the following elementary process:*

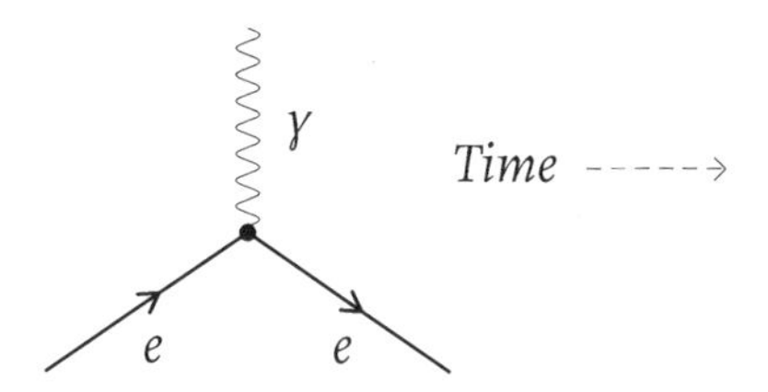

> *In these figures time flows horizontally, to the right, so this diagram reads: a charged particle, e, enters, emits (or absorbs) a photon, γ, and exits. For the sake of argument, I'll assume that the charged particle is an electron; it could just as well be a quark, or any lepton except a neutrino (the latter is neutral, of course, and does not experience an electromagnetic force)."*

We find here a series of misconceptions of which I will only comment on a few. The first misconception is to claim that quantum electrodynamics is a dynamical theory, when it is not. The t that appears in its equations is not time, but a formal label with no physical meaning. The same goes for the symbol x, which is not position in a physical space. The direct consequence of all this is that quantum electrodynamics cannot predict any dynamic phenomena, only describe simple scattering processes, because this theory is built on a fictitious spacetime:[70]

> *"The question arises whether it is natural to incorporate directly the classical Minkowski space-time in quantum field theory.* [...] *Every physicist would easily convince himself that all quantum calculations in particle physics are made in the energy-momentum space and that the Minkowski x^{μ} are just dummy variables without physical meaning (although almost all textbooks insist on the fact that these variables are not related with position, they use them to express locality of interactions!)."*

Griffiths' claim that any electromagnetic phenomena can be reduced to the simple diagram reproduced on the previous page is another common misconception, because quantum electrodynamics cannot describe arbitrary electromagnetic phenomena in bound states. Quantum electrodynamics cannot even describe the ancient Coulomb interaction!

The third and last misconception I want to mention is that Griffiths interprets the diagram in his book as if it represented a physical process. Nothing could be further from the truth! What he drawn is not an "*elementary process*". I just mentioned above that the diagram is associated to a fictitious spacetime, now I must add that e and γ are not true particles. The e do not represent electrons, but mathematical objects whose masses and charges do not correspond to anything found in nature. It is during the renormalization procedure that those unphysical e that appear in the diagrams and equations of quantum field theory are systematically replaced by the electrons that we observe in the experiments. Moreover, even after renormalization, γ cannot represent any particle since the diagram clearly shows that γ would have to travel a finite 'distance' in zero 'time'. No real particle can travel faster than light. γ in that diagram is what Feynman called a virtual photon, and as the name implies, a virtual particle cannot be real. There are more misconceptions in that brief excerpt taken from Griffith's book on elementary particles, but I will stop here.

The main **conclusions** of this chapter are the following:

- Electrodynamics in curved spacetime does not provide an unambiguous combination of electromagnetism and gravitation.
- In general relativity not even the energy of a single charge is well defined.
- Thermoelectric phenomena like the Seebeck effect are ignored in the Lorentz force law. Contrary to what is stated in electrodynamics textbooks, the Maxwell equations combined with the Lorentz force equation do not form "*the basis of all classical electromagnetic phenomena*".
- Contrary to myth, classical electrodynamics is not a limit of quantum electrodynamics.
- The electroweak theory does not provide a real unification of weak and electromagnetic interactions.
- The electromagnetic field is not fundamental at the quantum level, nor can vacuum fluctuations explain phenomena like the Casimir forces.
- Contrary to a widespread misconception, Feynman diagrams do not represent any real processes, nor can all electromagnetic phenomena be reduced to elementary diagrams. The e and γ that appear in these diagrams do not represent any known particles since they have unphysical masses, charges, and speeds.

Appendix A

THE ELECTROMAGNETIC FIELD FROM ACTION-AT-A-DISTANCE INTERACTIONS

Our starting point is the following action-at-a-distance Hamiltonian for a system of N interacting charges

$$H = \sum_{i}^{N} H_i + U, \tag{A.1}$$

where $H_i = \sqrt{m_i^2 c^4 + \boldsymbol{p}_i^2 c^2}$ are free terms and U the interaction between the charges. The general form of the interaction in terms of positions and momenta is quite complex, but with a very good level of accuracy it can be approximated by

$$U = \frac{1}{8\pi\epsilon_0} \sum_{i}^{N} \sum_{j \neq i}^{N} \int \frac{q_i \boldsymbol{p}_i^{\alpha} c}{H_i} D_{ij} \frac{q_j \boldsymbol{p}_j^{\beta} c}{H_j} \eta_{\alpha\beta}\, \mathrm{d}t' . \tag{A.2}$$

Here, D_{ij} is a phase space function that generates the Maxwell equations in potential form. The condition $j \neq i$ implies that self-interactions are excluded and that the whole Hamiltonian is finite, unlike in field theory, where the Hamiltonian (2.18) for a system of charges and field includes infinities that must be removed before making contact with the real world. Even at the risk of being annoying, we must quote Dirac once again: "*the theory has to be based on sound mathematics, in which one neglects only quantities that are small. One is not allowed to neglect infinitely large quantities*".[44]

Instead of working with the general form of (A.1), we will work with an approximate Hamiltonian where the square roots in the free terms are expanded into a power series in the inverse of the speed of light

$$\sqrt{m_i^2 c^4 + \boldsymbol{p}_i^2 c^2} = m_i c^2 + \frac{\boldsymbol{p}_i^2}{2m_i} + \mathcal{O}(c^{-2}). \tag{A.3}$$

The rest-energy terms $m_i c^2$ can be ignored, since they are mere constants; therefore, we will approximate the square roots with the quadratic terms $\boldsymbol{p}_i^2/2m_i$ characteristic of the Newton-Coulomb theories. U is also expanded in a power series, but terms up to c^{-2} order are preserved. The result is the following Hamiltonian

$$H = \sum_i^N \frac{\boldsymbol{p}_i^2}{2m_i} + \sum_i^N \sum_{j\neq i}^N \frac{q_i q_j}{8\pi\epsilon_0 r_{ij}} \left[1 - \frac{\boldsymbol{p}_i \cdot \boldsymbol{p}_j}{2m_i m_j c^2} - \frac{(\boldsymbol{p}_i \cdot \boldsymbol{r}_{ij})(\boldsymbol{p}_j \cdot \boldsymbol{r}_{ij})}{2m_i m_j c^2 r_{ij}^2} \right]. \tag{A.4}$$

This approximation to (A.1) greatly simplifies the calculations, but still allows us to retain the main physical insights to understand the nature of the electromagnetic field.

Note: Some readers might complain that the above expansion is not fully consistent because we retain terms up to c^{-2} order in the interaction part of the Hamiltonian, but we neglect terms up to the same order in the free parts. However, our goal is to derive the low-velocity Hamiltonian used in field theory (see equation 9.37 in the text[9] by Schwinger et al.), and field theory has that kind of inconsistency.

The Hamiltonian (A.4) can be used in practical applications as is, but to make contact with field theory, we must introduce the following *particle potentials*

$$\Phi_i = \sum_{j \neq i} \frac{q_j}{4\pi\epsilon_0 r_{ij}} \tag{A.5}$$

and

$$\boldsymbol{A}_i = \sum_{j \neq i} \frac{q_j}{8\pi\epsilon_0 m_j c^2 r_{ij}} \left[\boldsymbol{p}_j + \frac{\boldsymbol{r}_{ij}(\boldsymbol{p}_j \cdot \boldsymbol{r}_{ij})}{r_{ij}^2} \right], \tag{A.6}$$

obtaining a more concise expression for the Hamiltonian

$$H = \frac{1}{2} \sum_{i}^{N} \left[\frac{\boldsymbol{p}_i^2}{m_i} + q_i \Phi_i - \frac{q_i}{m_i} \boldsymbol{p}_i \cdot \boldsymbol{A}_i \right]. \tag{A.7}$$

Those particle potentials are simply abbreviations to simplify some expressions and have no special physical interpretation.

The above Hamiltonian is defined in a $6N$ phase space $(\boldsymbol{r}^{\{N\}}, \boldsymbol{p}^{\{N\}})$. However, field theory is formulated in a $4D$ spacetime and, as a first consequence, velocities acquire a more relevant status than momenta in the description of interactions. In a first step to derive the electromagnetic field from action-at-a-distance interactions, we need to replace momenta with velocities.

We can obtain the relationship between velocities and momenta from the first set of Hamiltonian equations

$$\boldsymbol{v}_i = \left(\frac{\partial H}{\partial \boldsymbol{p}_i} \right)_{\boldsymbol{r}} = \frac{\boldsymbol{p}_i}{m_i} - \sum_{j \neq i}^{N} \frac{q_i q_j}{8\pi\epsilon_0 r_{ij}} \left[\frac{\boldsymbol{p}_j}{m_i m_j c^2} + \frac{\boldsymbol{r}_{ij}(\boldsymbol{p}_j \cdot \boldsymbol{r}_{ij})}{m_i m_j c^2 r_{ij}^2} \right] = \frac{\boldsymbol{p}_i}{m_i} - \frac{q_i}{m_i} \boldsymbol{A}_i, \tag{A.8}$$

which allows us to rewrite the Hamiltonian in terms of velocities

$$H = \frac{1}{2} \sum_{i}^{N} \left[m_i \boldsymbol{v}_i^2 + q_i \Phi_i + q_i \boldsymbol{v}_i \cdot \boldsymbol{A}_i \right]. \tag{A.9}$$

Despite its form, this is still a true Hamiltonian defined in phase space, because the velocities $\boldsymbol{v}_i = \boldsymbol{v}_i(\{\boldsymbol{p}_i\}^N, \{\boldsymbol{r}_i\}^N)$ are function of momenta and positions and still $H = H(\{\boldsymbol{p}_i\}^N, \{\boldsymbol{r}_i\}^N)$. This Hamiltonian depends on $2N$ particle potentials, while field theory uses only a pair of potentials, Φ and $\boldsymbol{A}$. We call these *field potentials*. The action-at-a-distance approach uses $2N$ potentials because it contains more information about the interactions than the field approach. The missing information is responsible for all the deficiencies and paradoxes of the field theory of electromagnetism.

The connection with the potentials of the field theory is achieved by first removing the physical constraint $j \neq i$ and then introducing correction terms. For the scalar particle potential,

$$\Phi_i = \sum_{j\neq i}^{N} \frac{q_j}{4\pi\epsilon_0 r_{ij}} = \underbrace{\sum_{j}^{N} \frac{q_j}{4\pi\epsilon_0 r_{ij}}}_{\Phi} - \underbrace{\frac{q_i}{4\pi\epsilon_0 r_{ii}}}_{\Phi_i^{\text{self}}} . \tag{A.10}$$

The first term on the right is common for all the particles because the summation no longer depends on i, and we call it Φ. The last term represents a *hypothetical* interaction Φ_i^{self} of the charge with itself. The net result is that we can write the scalar particle potential as $\Phi_i = (\Phi - \Phi_i^{\text{self}})$, and can write a similar expression $\boldsymbol{A}_i = (\boldsymbol{A} - \boldsymbol{A}_i^{\text{self}})$ for the vector particle potential. The Hamiltonian is now

$$H = \frac{1}{2}\sum_{i}^{N}\left[m_i\boldsymbol{v}_i^2 + q_i\left(\Phi - \Phi_i^{\text{self}}\right) + q_i\boldsymbol{v}_i\cdot\left(\boldsymbol{A} - \boldsymbol{A}_i^{\text{self}}\right)\right]. \tag{A.11}$$

The new potentials Φ and $\boldsymbol{A}$ diverge because $r_{ij} = 0$ when $j = i$, but the corrections Φ_i^{self} and $\boldsymbol{A}_i^{\text{self}}$ eliminate this divergence, which makes the Hamiltonian (A.11) still a finite quantity. We will see in a moment that the absence of these correction terms in field theory is the reason why field theory predicts meaningless infinite energies.

The above expression for the Hamiltonian can be rearranged as

$$H = \frac{1}{2}\sum_{i}^{N}(m_i \boldsymbol{v}_i^2 - q_i \boldsymbol{v}_i \cdot \boldsymbol{A}_i^{\text{self}}) + \frac{1}{2}\sum_{i}^{N}(q_i \Phi + q_i \boldsymbol{v}_i \cdot \boldsymbol{A}) + C, \quad \text{(A.12)}$$

where $C = -(1/2)\sum_i^N q_i \Phi_i^{\text{self}}$. Multiplying both sides of (A.8) with m_i, replacing $\boldsymbol{A}_i$ by $(\boldsymbol{A} - \boldsymbol{A}_i^{\text{self}})$ and rearranging again the result gives

$$(m_i \boldsymbol{v}_i - q_i \boldsymbol{A}_i^{\text{self}}) = (\boldsymbol{p}_i - q_i \boldsymbol{A}). \quad \text{(A.13)}$$

Multiplying both sides of this result by $\boldsymbol{v}_i$ and iterating, finally produces the following identity

$$(m_i \boldsymbol{v}_i^2 - q_i \boldsymbol{v}_i \cdot \boldsymbol{A}_i^{\text{self}}) = (\boldsymbol{p}_i - q_i \boldsymbol{A})\frac{1}{m_i - m_i^{\text{self}}}(\boldsymbol{p}_i - q_i \boldsymbol{A}), \quad \text{(A.14)}$$

where $m_i^{\text{self}} = (q_i/\boldsymbol{v}_i)\boldsymbol{A}_i^{\text{self}}$. This is a divergent quantity called the *self-mass* or the *electromagnetic mass* in the field theory literature. This is essentially equivalent to the δm that we use in chapter 7 for a single charge.

Replacing $\boldsymbol{A}_i^{\text{self}}$ by its value, expanding it in a power series in $(\boldsymbol{v}/c)$ and retaining only the first term of the series (which is equivalent to evaluate it in the rest frame of the particle), we obtain for the self-mass

$$m_i^{\text{self}} = \frac{q_i^2}{4\pi\epsilon_0 r_{ii} c^2}, \quad \text{(A.15)}$$

a result first obtained by Dirac for stationary regimes. Note that the specific value of the self-mass for arbitrary frames or dynamical regimes is irrelevant since it is an nonphysical quantity and cannot be observed.

The identity (A.14) can be used in (A.12) to rewrite the Hamiltonian

$$H = \frac{1}{2}\sum_{i}^{N}\frac{(\boldsymbol{p}_i - q_i \boldsymbol{A})^2}{m_i - m_i^{\text{self}}} + \frac{1}{2}\sum_{i}^{N}(q_i \Phi + q_i \boldsymbol{v}_i \cdot \boldsymbol{A}) + C. \quad \text{(A.16)}$$

If we now combine real masses and self-masses into a new formal concept of *bare mass* defined by $m_i^{\mathrm{bare}} = (m_i - m_i^{\mathrm{self}})$, and then use $\rho(\boldsymbol{r}, t) = \sum_i^N q_i \delta(\boldsymbol{r} - \boldsymbol{r}_i(t))$ and $\boldsymbol{J}(\boldsymbol{r}, t) = \sum_i^N q_i \mathbf{v}_i \delta(\boldsymbol{r} - \boldsymbol{r}_i(t))$, we finally obtain

$$H = \sum_i^N \frac{(\boldsymbol{p}_i - q_i \mathbf{A})^2}{2m_i^{\mathrm{bare}}} + \frac{1}{2} \int (\rho \Phi + \boldsymbol{J} \cdot \mathbf{A}) \, \mathrm{d}V + C. \tag{A.17}$$

The integral can be written in an alternative form,[71] obtaining the standard result as function of the electric and magnetic components of the electromagnetic field

$$H = \sum_i^N \frac{(\boldsymbol{p}_i - q_i \mathbf{A})^2}{2m_i^{\mathrm{bare}}} + \epsilon_0 \int \frac{\mathbf{E}^2 + c^2 \mathbf{B}^2}{2} \, \mathrm{d}V + C. \tag{A.18}$$

Except by the renormalization term C, this Hamiltonian is formally analogous to the one proposed by field theory (2.19). We can affirm that the action-at-a-distance theory is formally equivalent to renormalized field theory.

Using a Legendre transformation, we can obtain the Lagrangian corresponding to (A.18)

$$L = \sum_i^N \frac{1}{2} m_i^{\mathrm{bare}} \mathbf{v}_i^2 + \int (\boldsymbol{J} \cdot \mathbf{A} - \rho \Phi) \, \mathrm{d}V + \epsilon_0 \int \frac{\mathbf{E}^2 - c^2 \mathbf{B}^2}{2} \, \mathrm{d}V - C, \tag{A.19}$$

again showing that the action-at-a-distance approach in terms of physical particles is formally equivalent to a field theory description in terms of bare particles plus renormalization counterterms.

We have made it notationally explicit in (A.18) and (A.19) that the masses used in field theory are not the experimental masses of the charges, but the fictitious quantities m_i^{bare}. The mainstream literature is rather opaque in this regard. Field theorists write m_i in the initial equations, and only when self-actions are considered is it revealed to the reader that m_i is not the true mass of the charge, but that the true mass is $(m_i + \delta m_i)$, where δm_i is a renormalization correction. We saw an example of this attitude in chapter 7, during the field-theoretic derivation of the Abraham-Lorentz equation (7.24).

It would be didactically preferable for field theorists to write m_i^{bare} in the initial equations, and m_i in the final renormalized equations. Our approach does not have these shortcomings since we start with an action-at-distance theory that uses the real masses of the charged particles and there is no need to renormalize the resulting equations of motion.

The correction term C keeps our Hamiltonian (A.18) finite and physical, unlike the field theory Hamiltonian, which lacks this term. We can obtain further insight about this correction term if we split the electric component of the electromagnetic field into longitudinal and transverse components, $\boldsymbol{E}^2 = \boldsymbol{E}_{\mathrm{L}}^2 + \boldsymbol{E}_{\mathrm{T}}^2$, because we can see that

$$\int \frac{\epsilon_0 \boldsymbol{E}_{\mathrm{L}}^2}{2} \, \mathrm{d}V + C = \sum_i^N \sum_{j \neq i}^N \frac{q_i q_j}{8\pi\epsilon_0 r_{ij}}, \tag{A.20}$$

which implies that C ensures that the longitudinal component provides the correct Coulomb interaction between the charges.

Since the field theory Hamiltonian is missing the C term, physicists simply replace the integral of the longitudinal component by the Coulomb interaction (see, for instance equation 1.57 in the textbook[65] by Mandl and Shaw)

$$\int \frac{\epsilon_0 \boldsymbol{E}_{\mathrm{L}}^2}{2} \, \mathrm{d}V \longrightarrow \sum_i^N \sum_{j \neq i}^N \frac{q_i q_j}{8\pi\epsilon_0 r_{ij}}, \tag{A.21}$$

with the two authors highlighting that "*we have dropped the infinite self-energy which occurs for point charges*".

Not only can we not drop terms that we do not like, but we also cannot neglect infinitely large quantities and keep smaller quantities. Any mathematician would faint if he discovered that field physicists approximate $(5 + \infty)$ by 5.

Those difficulties are not present in the action-at-a-distance approach because the C term in (A.18) cancels out the infinite self-energy from the field, ensuring a finite energy. The rest of the infinities cancel out thanks to the difference between real masses and bare masses.

Mandl and Shaw started with the field expression for the energy and wanted to 'derive' the energy of a system of charges *after dropping an unwanted infinity*. Greiner and Jackson do the opposite[6,8]

$$\sum_{i}^{N}\sum_{j\neq i}^{N}\frac{q_i q_j}{8\pi\epsilon_0 r_{ij}} \longrightarrow \int\frac{\epsilon_0 \mathbf{E}_{\mathrm{L}}^2}{2}\,\mathrm{d}V. \tag{A.22}$$

Both authors start with the Coulomb expression for the energy, replace the charges with continuous charge distributions, and use the Poisson equation to finally obtain an expression in terms of the electric component of the field. Both authors also consider an empirical situation in which $\mathbf{E}^2 = \mathbf{E}_{\mathrm{L}}^2$, but this does not change the main point that they are adding an unwanted infinity to the energy in order to obtain the field expression.

Unlike Jackson, Greiner warns his readers that when "*translating the sum (1.103) to an integral (1.104), we have not taken into account the condition i ≠ j appearing in (1.103) [...] so that (1.104) contains automatically self-energy parts which become infinitely large for point charges*". However, he does not tell readers that these infinities must later be eliminated to come into contact with observations and experiments.

Greiner and Jackson interpret the quantity $(\epsilon_0 \mathbf{E}_\mathrm{L}^2/2)$ within the integral as the *"energy density of the electric field"*, although Greiner uses a different system of units. Now, Jackson makes a very important comment about how the potential energy of two charges of opposite sign is negative, while the integral term is positive since the density cannot be negative. Jackson also adds that *"the reason for this apparent contradiction is that (1.54) and (1.55) contain 'self-energy' contributions to the energy density, whereas the double sum in (1.51) does not"*. Indeed, positive infinity minus any finite quantity is positive infinity, but like Greiner, Jackson also fails to tell us that such infinities must be removed in the end and that the actual measured energy of two charges of opposite sign is negative.

The use of bare masses and the lack of the renormalization term are the reasons of why the field theory of electromagnetism is an inconsistent theory that gives unphysical infinities and violates the laws of conservation and inertia. We saw in chapter 4 that physicists disagree on how to physically interpret each term in the field theory Hamiltonian. They disagree about what the energy of matter is and what the energy of the field is, and now we can understand why. There is no electromagnetic field, there are only directly interacting particles, so none of the terms of the Hamiltonian can be consistently and unambiguously identified with a field. Specifically, we can see that neither the expression

$$\epsilon_0 \int \left(\frac{\mathbf{E}^2 + c^2 \mathbf{B}^2}{2} - \frac{\rho \Phi}{\epsilon_0} \right) \mathrm{d}V \tag{A.23}$$

in the Hamiltonian (4.5) nor

$$\epsilon_0 \int \left(\frac{\mathbf{E}^2 + c^2 \mathbf{B}^2}{2} - \Phi \nabla \cdot \mathbf{E} \right) \mathrm{d}V \tag{A.24}$$

in the Hamiltonian (4.6) represent the energy of a *"free field"*, since both are derived from $(1/2) \sum_i^N (q_i \Phi_i + q_i \mathbf{v}_i \cdot \mathbf{A}_i)$ for the charges.

Using again the decomposition of the electric part of the field into longitudinal and transverse components, $\boldsymbol{E}^2 = \boldsymbol{E}_\mathrm{L}^2 + \boldsymbol{E}_\mathrm{T}^2$, we can rewrite the action-at-a-distance Hamiltonian (A.18) as

$$H = \sum_i^N \frac{(\boldsymbol{p}_i - q_i\boldsymbol{A})^2}{2m_i^{\mathrm{bare}}} + \sum_i^N \sum_{j \neq i}^N \frac{q_i q_j}{8\pi\epsilon_0 r_{ij}} + \epsilon_0 \int \frac{\boldsymbol{E}_\mathrm{T}^2 + c^2\boldsymbol{B}^2}{2} \mathrm{d}V. \quad \text{(A.25)}$$

Except for using a different system of units, the integral term is what Mandl and Shaw associate with the energy of radiation and denote by H_{rad}. Their interpretation is untenable because we have just derived the same term after assuming there is no radiation, only charged matter. Our development confirms that the usual interpretation of the field theory of electromagnetism does not hold up on closer examination. This is another reason why field theory is an inconsistent theory. Thus, the theory not only suffers from mathematical difficulties such as divergences, but also lacks a coherent (physical) interpretation.

The conclusion we reach from the developments in this appendix is that the action-at-a-distance approach to electromagnetic interactions is a superior approach because it provides physical energies, along with finite Hamiltonians and Lagrangians, and also explains why field theory uses nonphysical bare parameters instead of the real (measured) masses of the charged particles. As we advanced during the previous chapters of this book, *the action-at-a-distance approach has not problem with the concept of point-like particles*, only field theory has this problem and requires a renormalization procedure.

The physical interactions between charges can be described by a set of N scalar and vector potentials that, in general, take different values for each particle, $\Phi_i \neq \Phi_j$ and $\boldsymbol{A}_i \neq \boldsymbol{A}_j$. However, field theory assumes an idealized symmetry, a "*principle of spacetime uniformity*" in Schwinger's words, where nothing in the description of the interactions would distinguish one charge from another, apart from the reference to the region of spacetime that each one occupies. This idealized symmetry forces interactions to be described in terms of generic potentials Φ and $\boldsymbol{A}$ common to all the charges, which introduces nonphysical interaction modes that must be absorbed in the bare masses m_i^{bare} and in a renormalization term C.

Note: In addition to bare masses, quantum field theory also relies on bare charges q_i^{bare}; if we had made a full quantum treatment of the electromagnetic interaction, we would also explain the presence of bare charges in quantum field theory.

We have seen throughout this book that the field theory of electromagnetism cannot fully describe the motion of a system of interacting charges. The main reason is that field theory contains infinite degrees of freedom and cannot provide a closed set of equations for a finite number of charges. Even if this problem were somehow solved, field theory still could not describe the motion of a system of interacting charges since the terms $\sum_i^N [(\boldsymbol{p}_i - q_i\boldsymbol{A})^2/2m_i^{\text{bare}} + q_i\Phi]$ associated to the charges in field theory cannot be used to derive the N-body equations of motion because part of the charge-charge interaction is hidden in what field theorists call the free field, as shown above.

Appendix B

SPEED OF MASSLESS BOSONS

We saw on pages 47 and 48 that a very common misconception found in electrodynamics textbooks is the statement that c is "*the velocity of light*". We then explained that c is a speed since it is a scalar and not a vector. Now we will go one step further and show in this appendix why photons cannot have velocity.

We start with the Hamiltonian for a free particle

$$H = \sqrt{m^2c^4 + \boldsymbol{p}^2c^2}. \tag{B.1}$$

Using the first Hamilton equation, we obtain the velocity of the particle

$$\boldsymbol{v} = \left(\frac{\partial H}{\partial \boldsymbol{p}}\right)_r = \frac{\boldsymbol{p}c^2}{\sqrt{m^2c^4 + \boldsymbol{p}^2c^2}}. \tag{B.2}$$

For a massless particle like the photon, this expression reduces to

$$\boldsymbol{v} = \frac{\boldsymbol{p}}{|\boldsymbol{p}|}c. \tag{B.3}$$

It seems that we have obtained an expression for the velocity of a photon, and confirmed that c is the speed, $|\boldsymbol{v}| = c$, but in reality (B.3) is not valid, as we will see in a moment.

Massive particles can be accelerated and decelerated. Indeed, if we replace $\boldsymbol{p} \to \lambda \boldsymbol{p}$ in (B.2), then the velocity changes according to the following rule

$$\frac{\boldsymbol{p}c^2}{\sqrt{m^2c^4 + \boldsymbol{p}^2c^2}} \longrightarrow \frac{\lambda}{|\lambda|} \frac{\boldsymbol{p}c^2}{\sqrt{\frac{m^2c^4}{\lambda^2} + \boldsymbol{p}^2c^2}} . \tag{B.4}$$

Note: Mathematically, $\sqrt{\lambda^2} = \pm|\lambda|$. However, we will only consider positive energies and square roots expressions like $\sqrt{m^2c^4 + \boldsymbol{p}^2c^2}$ must be taken in the mathematical sense of $\left|\sqrt{m^2c^4 + \boldsymbol{p}^2c^2}\right|$.

To continue our analysis, we can consider an experimental situation in which the initial velocity of the particle is positive and is first reduced by half, then reduced to zero, and finally reduced to minus the initial value,

$$+\boldsymbol{v} \longrightarrow +\tfrac{1}{2}\boldsymbol{v} \longrightarrow \boldsymbol{0} \longrightarrow -\boldsymbol{v} . \tag{B.5}$$

This is what happens, for example, when a charged particle is repelled by an electric force antiparallel to the initial motion. In absence of dissipation, the velocity of the particle takes all the possible values between $+\boldsymbol{v}$ and $-\boldsymbol{v}$, and to simplify our discussion, we are considering only two of those intermediate values.

If we apply the same transformation $\boldsymbol{p} \to \lambda \boldsymbol{p}$ to (B.3), we obtain that the velocity of a massless particle changes as

$$\frac{\boldsymbol{p}}{|\boldsymbol{p}|}c \longrightarrow \frac{\lambda}{|\lambda|}\frac{\boldsymbol{p}}{|\boldsymbol{p}|}c , \tag{B.6}$$

and from this we obtain the surprising results that $\boldsymbol{v}$ does not change when $0 < \lambda$, it does change when $\lambda < 0$, and it becomes undefined when $\lambda = 0$, because $(0/0)$ is an indeterminate form in mathematics.

Note: Why is the form $(0/0)$ undefined? Well, $(z/z) = 1$ for a nonzero number z because the division is the inverse of the multiplication, and if we multiply both sides of the equation $(z/z) = 1$ by an arbitrary number n, we obtain the expected result $(n \cdot z)/z = n$. Now, if we *assume* that $(0/0) = 1$ and multiply both sides by n, we obtain $(n \cdot 0)/0 = (0/0) = n$. As a result, we obtain that the factor $(0/0)$ is equal to any arbitrary number n. The 'solution' to all this trouble is that the form $(0/0)$ is undefined.

Considering the cases $\lambda = 1/2$, $\lambda - 1$, and $\lambda = 0$ in (B.6), we confront to the next mystery: How could the velocity of a photon moving with speed c be reduced from $+\boldsymbol{v}$ to $-\boldsymbol{v}$ when it turns out that a photon cannot be decelerated to half the original velocity, and a zero velocity cannot be even defined for it?

$$+\boldsymbol{v} \not\longrightarrow +\tfrac{1}{2}\boldsymbol{v} \not\longrightarrow \frac{\mathbf{0}}{|\mathbf{0}|}c \not\longrightarrow -\boldsymbol{v}\,. \qquad (+\boldsymbol{v} \overset{?}{\longrightarrow} -\boldsymbol{v}) \tag{B.7}$$

We can obtain all kind of weird paradoxes similar to these, and the ultimate reason is that the expression (B.3) for the velocity is not valid. We have made a very subtle mistake. The Hamiltonian (B.1) is a function of position and momentum $H = H(\boldsymbol{r}, \boldsymbol{p})$. Of course, $(\partial H/\partial \boldsymbol{r})_{\boldsymbol{p}} = 0$, but the particle must have position in order to be able to obtain its velocity from $(\mathrm{d}\boldsymbol{r}/\mathrm{d}t) = (\partial H/\partial \boldsymbol{p})_{\boldsymbol{r}}$. However, massless particles have no position and the Hamilton equation (B.2) is only valid when $m \neq 0$. For massless particles, neither the derivative $(\mathrm{d}\boldsymbol{r}/\mathrm{d}t)$, nor the partial derivative $(\partial H/\partial \boldsymbol{p})_{\boldsymbol{r}}$ are defined and (B.3) cannot be used.

The demonstration of why massless particles cannot be localized is rather technical and will be given in another volume of this series. Suffice to say now that we can define a rest frame for massive particles as the frame in which the momentum of the particle is zero, because when we set $\boldsymbol{p} = \mathbf{0}$ in (B.2), we obtain that the particle is at rest, since $\boldsymbol{v} = \mathbf{0}$. However, massless particles *always* have a speed c, and there is no rest frame for them. The impossibility of finding a frame where massless particles are at rest is complementary to the absence of a concept of position for them.

Appendix C

LIST OF MISCONCEPTIONS

This is a list of the common misconceptions mentioned in the book, categorized by chapter. So, for example, when an entry in this list is "*Classical electrodynamics is a field theory*", this means that this is a common misconception, because, as explained in the corresponding chapter, there is a alternative formulation of classical electrodynamics in terms of action at a distance, which does not use fields.

INTRODUCTION

1) Biophysics is a true unification of biology and physics.
2) Quantum chemistry unifies physics and chemistry.
3) The classical theory of electromagnetism is a satisfactory theory all by itself.
4) When electromagnetism is combined with quantum mechanics, the difficulties of the classical theory are resolved.

THE THEORETICAL MINIMUM

5) Electrodynamics is a beautifully complete and successful theory.

6) The formalism of electrodynamics is an ideal model that other theories strive to emulate.
7) The vacuum in classical field theory is a region of space devoid of any system.
8) There are almost independent electric and magnetic fields.
9) We must talk about electromagnetic fields.
10) $\nabla \cdot \boldsymbol{A}^{\mathrm{Lor}} + 1/c^2\,(\partial\Phi^{\mathrm{Lor}}/\partial t) = 0$ is the Lorentz gauge.
11) The four Maxwell equations together with the Lorentz force law summarize the entire theoretical content of classical electrodynamics, except for some special properties of matter.
12) When combined with the Lorentz force equation and Newton's second law of motion, the Maxwell equations provide a complete description of the classical dynamics of interacting charged particles and electromagnetic fields.
13) The space around an electric charge is permeated by electric and magnetic fields.
14) Fields mediate the interaction between charges by transmitting the influence from one charge to the other.
15) Electromagnetic fields can exist in regions of space where there are no sources.

ACTION AT A DISTANCE

16) Newton was extremely unhappy with action at a distance.
17) In the action-at-a-distance model, causality arises as an influence that propagates with a finite speed between isolated particles
18) Classical electrodynamics is a field theory.
19) Experiments imply that fields are fundamental.
20) Field theory explains how charges interact.

21) Hertz's experimental work confirmed the field concept.

22) The contact-action model solved the mysteries of the action-at-a-distance model.

23) The scalar potential $\Phi(\boldsymbol{r}, t)$ obtained from the Maxwell equations is just the Coulomb potential.

LAGRANGIANS AND HAMILTONIANS

24) The Lagrangian $-\sum_i^N m_i c^2 \sqrt{1 - \mathbf{v}_i^2/c^2}$ describes free charges.

25) The Lagrangian $\epsilon_0 \int (\boldsymbol{E}^2 - c^2 \boldsymbol{B}^2)/2 \, \mathrm{d}V$ describes a free field.

26) The Lagrangian $- \sum_i^N m_i c^2 \sqrt{1 - \mathbf{v}_i^2/c^2} + \int (\boldsymbol{J} \cdot \boldsymbol{A} - \rho \Phi) \, \mathrm{d}V$ describes a system of charges.

27) The Lagrangian $\int (\boldsymbol{J} \cdot \boldsymbol{A} - \rho \Phi) \, \mathrm{d}V + \epsilon_0 \int (\boldsymbol{E}^2 - c^2 \boldsymbol{B}^2)/2 \, \mathrm{d}V$ describes the electromagnetic field.

28) The energy of free charges is $\sum_i^N \sqrt{m_i^2 c^4 + (\boldsymbol{p}_i - q_i \boldsymbol{A})^2 c^2}$.

29) $\epsilon_0 \int (\boldsymbol{E}^2 + c^2 \boldsymbol{B}^2)/2 \, \mathrm{d}V$ gives the energy of the electromagnetic field.

30) The energy of a free field is $\epsilon_0 \int [(\boldsymbol{E}^2 + c^2 \boldsymbol{B}^2)/2 - \rho \Phi / \epsilon_0] \mathrm{d}V$.

31) $\sum_i^N \sqrt{m_i^2 c^4 + (\boldsymbol{p}_i - q_i \boldsymbol{A})^2 c^2} + q_i \Phi$ gives the energy of a system of interacting charges.

32) The energy of a free field is $\epsilon_0 \int [(\boldsymbol{E}^2 + c^2 \boldsymbol{B}^2)/2 - \Phi \nabla \cdot \boldsymbol{E}] \mathrm{d}V$.

33) $\sum_i^N q_i \Phi$ is the potential energy of a system of charges.

34) Fields are physical systems in their own right, every bit as 'real' as atoms or baseballs.

DELAYED INTERACTIONS

35) Experiments show that instantaneous interactions do not exist in nature.

36) The principle of causality for perturbations propagating into the future is only taken into account by the retarded potentials.

37) $c = 299792458$ m/s, a scalar quantity, is the velocity of light in empty space.

38) Quantities calculated from iteration procedures are not physical.

39) The Maxwell displacement current is not a true current.

40) The Coulomb gauge is not a physical gauge.

41) The only viable way to calculate electric and magnetic fields in terms of their causative sources is to use retarded field integrals (or retarded potentials).

42) A mechanics based on the assumption of an instantaneous propagation of interactions contains in itself a certain inaccuracy.

43) Instantaneous potentials violates special relativity and the principle of causality.

44) The Coulomb gauge has potentials with peculiar 'unphysical' relativistic properties.

45) The Maxwell equations do not represent an instantaneous action-at-a-distance theory.

46) Field theory and action-at-a-distance are formally equivalent.

47) Although the advanced potentials are consistent with Maxwell equations, they violate the most sacred tenet in all of physics: the principle of causality.

48) Although the advanced potentials are of some theoretical interest, they have no direct physical significance.

49) The law of action and reaction is not valid in a special-relativistic context.

50) The advanced potential represents an electromagnetic wave traveling backward in time.

ELECTROMAGNETIC FORCES

51) The Lorentz force is the force that acts on a point charge q in the presence of an electromagnetic field.

52) The Lorentz force is the total electromagnetic force on a charged particle.

53) The momentum of a charged particle is $m\gamma\boldsymbol{v}$.

54) The Lorentz force gives the rate of change of the particle's momentum.

55) ($m\boldsymbol{a} = \boldsymbol{F}^{\mathrm{Lor}}$) is the equation of motion for the Lorentz force.

56) It is understood that $\boldsymbol{E}$ and $\boldsymbol{B}$ in the Lorentz force law $\boldsymbol{F}^{\mathrm{Lor}} = q(\boldsymbol{E} + \boldsymbol{v} \times \boldsymbol{B})$ are external fields.

RADIATION REACTION AND RENORMALIZATION

57) The total momentum for the system of a charged particle and the electromagnetic field is $(m\gamma\boldsymbol{v} + \mathbf{G})$, where $\mathbf{G}$ is the total electromagnetic momentum of the field.

58) The particle's momentum is partly mechanical, but with an electromagnetic contribution.

59) Accelerated charges radiate.

60) The Abraham-Lorentz equation takes into account the radiative energy loss and its effect on the motion of the particle.

61) The Landau-Lifshitz equation is a sensible alternative to the Abraham-Lorentz equation for the classical regime of small radiative effects.

62) A point particle cannot be taken too literally in a classical context; it must always be considered as an approximation to a nonsingular and extended charge distribution.

63) Difference-delay equations remove runaways, preacceleration, and other pathologies.

64) The radiative potential, and only it, exerts a force on the particle.

65) The singular field does not exert a force on the particle (it merely contributes to the particle's inertia); the entire self-force arises from the action of the radiative field.

66) The masses that we measure in the laboratory are the combination of a theoretical negative infinite mass plus a positive infinite correction.

67) Infinities are an inevitable part of nature.

68) Field theory can produce a complete and satisfactory equation of motion for a single charged particle like the electron.

69) The motion of a charged particle can be completely described by the Abraham-Lorentz-Dirac equation.

70) Point particles cannot receive a completely consistent treatment in a classical theory of electromagnetism.

MANY-BODY MOTION

71) Field theory completely describes the motion of two or more interacting particles.

72) The big bang model used in cosmology introduced a beginning for time.

73) An infinite number of degrees of freedom is required to describe the motion of a system of interacting charges.

74) The Earth's magnetic field has been measured.

75) Relativistic field theory is essentially a one-body theory.

76) Only when the retardation effects can be neglected is a Lagrangian description possible in terms of instantaneous positions and velocities.

77) The Darwin Lagrangian neglects retardation effects.

78) We know that due to the finite velocity of propagation, the field must be considered as an independent system with its own 'degrees of freedom'.

79) It is impossible to rigorously describe the system of interacting particles with the aid of a Lagrangian, depending only on the coordinates and velocities of the particles and without containing quantities related to the internal 'degrees of freedom' of the field.

80) Wheeler and Feynman provided a fully satisfactory theory of electromagnetism.

THE COULOMB LIMIT

81) The scalar potential $\Phi(\boldsymbol{r}, t)$ in the Coulomb gauge is just the instantaneous Coulomb potential.

82) The SHP theory provides a complete and consistent many-body theory, valid for any speed.

83) There is a simultaneous and independent coexistence of Newton instantaneous long-range (NILI) and Faraday-Maxwell short-range interactions (FMSI), which cannot be reduced to each other.

84) $\boldsymbol{F}(\boldsymbol{r}) = q\boldsymbol{E}(\boldsymbol{r})$ is the Coulomb law.

85) Coulomb theory can be considered a limiting case of the Lorentz-Maxwell theory.

FLUCTUATIONS

86) The microscopic Maxwell equations govern electromagnetic phenomena in the microscopic world made up of electrons and nuclei.

87) Replacing the average values of the mechanical and electromagnetic quantities with the experimental stochastic values of these quantities is sufficient to correctly describe real-world electromagnetic phenomena.

88) The Maxwell-Lorentz equations are valid for any system where gravitation and quantum effects are negligible.

COVARIANT FORMULATION

89) u^α is a four-velocity because it contains $\boldsymbol{v}$, a three-velocity.

90) $K^\alpha = qF^{\alpha\beta}u_\beta$ is the electromagnetic Minkowski force.

91) The covariant formulation of electrodynamics provides a high degree of compactness.

92) The four-momentum p^α is a true four-vector in the sense that its components can vary freely.

93) The world is four-dimensional and the covariant quantities and equations do not have redundant degrees of freedom.

94) The quantity mu^α is the only reasonable candidate for a spacetime tensor that could represent the particle's momentum.

95) Only the covariant formulation provides a relativistic electrodynamics.

96) The mathematical equations expressing the laws of nature must be covariant, that is, invariant in form, under the transformations of the Lorentz group.

97) The Lorentz equation of motion has no well-defined status in special relativity and must be discarded and replaced by a new law involving only spacetime tensors.

98) Covariant Hamiltonians are true Hamiltonians.

99) $T^{\alpha\beta} = (1/\mu_0)[F^{\alpha\lambda}F^{\beta}_{\lambda} - (1/4)\eta^{\alpha\beta}F_{\lambda\zeta}F^{\lambda\zeta}]$ is the stress-energy-momentum tensor for the electromagnetic field.

THE NOETHER THEOREM

100) The Noether theorem is a general connection between symmetries and conserved quantities for physical systems.

101) Conservation laws follow from the symmetry properties of nature.

102) It is increasingly clear that the symmetry group of nature is the deepest thing that we understand about nature today.

103) What Noether called the 'laws of conservation' are always conservation laws.

104) $\partial_\alpha T^{\alpha\beta} = -F^{\alpha\beta}J_\alpha$ is a conservation law.

105) A quantity is conserved only when it is constant.

106) The law of conservation of energy, the first law of thermodynamics, is $\mathrm{d}E = 0$.

107) The first law of thermodynamics for any system is $\mathrm{d}E = đQ + đW$.

108) Energy is not conserved in dissipative systems.

109) Energy E and momentum $\boldsymbol{p}$ are conserved only when they are constants of the motion, but charge Q is conserved even when it is not constant.

MISCELLANY

110) General relativity is a theory of the gravitational field.

111) Electrodynamics in curved spacetime provides an unambiguous combination of electromagnetism and gravitation.

112) The energy of a single charge is well defined in general relativity.

113) The pseudo-Hamiltonian $1/2(p^\alpha - qA^\alpha)\,g_{\alpha\beta}\,(p^\beta - qA^\beta)$ describes the motion of a charged particle with charge q in a gravitational field.

114) The deviation $h_{\alpha\beta}$ from flatness is an observable that describes the gravitational field.

115) When the velocity of the charges is small enough, the Ohm law (also called the generalized Ohm law) gives the current density $\boldsymbol{J}$.

116) The production of entropy in a volume element due to electrical phenomena is given by $(\boldsymbol{J} \cdot \mathbf{E}/T)$.

117) Classical electrodynamics is a limit of quantum electrodynamics (for small momentum and energy transfers, and large average numbers of virtual or real photons).

118) The velocity of a charged particle is given in terms of Dirac alpha operators: $\hat{\boldsymbol{v}} = c\hat{\boldsymbol{\alpha}}$.

119) $q(\hat{\boldsymbol{E}} + c\hat{\boldsymbol{\alpha}} \times \hat{\boldsymbol{B}})$ is the quantum analogue of the Lorentz force.

120) The standard model provides a coherent quantum mechanical description of electromagnetic, weak, and strong interactions.

121) Relativistic quantum mechanics is known as quantum field theory.

122) The electroweak theory provides a real unification of weak and electromagnetic interactions.

123) The standard model brought together, in a single theoretical package, the weak interaction, the strong interaction, and electrodynamics.

124) The electromagnetic field together with the W and Z fields comprise a unified electroweak gauge field that describes both the electromagnetic and weak interactions.

125) String theory is the leading candidate for a theory of everything.

126) The classic idea that electromagnetic fields are produced by charges is particularly untenable in quantum field theory, since it is essential for the understanding of such phenomena as the vacuum fluctuations of the electromagnetic field that the electromagnetic field has its own dynamical degrees of freedom, independently of the existence of charged matter.

127) All electromagnetic phenomena are ultimately reducible to the elementary processes $e \to e + \gamma$ and $e + \gamma \to e$.

128) x and t in the formalism of quantum field theory represent space and time.

REFERENCES

1. Q&A: What is biophysics? **2002:** *BMC Biology 9(13), 1.* ZHOU, HUAN-XIANG.

2. Quantum Chemistry; fifth edition **2000:** *Prentice-Hall, Inc.; New Jersey.* LEVINE, IRA N.

3. Chemical Kinetics and Dynamics **2003:** *Annals of the New York Academy of Sciences 988, 128.* PRIGOGINE, ILYA.

4. The Feynman Lectures On Physics, Vol 2; Mainly Electromagnetism And Matter; Second Printing **1964:** *Addison-Wesley Publishing Company, Inc.; Reading, Massachusetts.* FEYNMAN, RICHARD P.; LEIGHTON, ROBERT B.; SANDS, MATTHEW.

5. Introduction to Electrodynamics; third edition **1999:** *Prentice-Hall, Inc.; New Jersey.* GRIFFITHS, DAVID J.

6. Classical Electrodynamics; third edition **1999:** *John Wiley & Sons, Inc.; New York.* JACKSON, J. D.

7. The Classical Theory of Fields; Third Revised English Edition **1971:** *In Course of Theoretical physics Volume 2; Butterworth-Heinemann; Amsterdan.* LANDAU, L. D.; LIFSHITZ E. M. (HAMMERMESH, M (TRANSLATOR)).

8 Classical electrodynamics (Classical theoretical physics) **1998:** *Springer-Verlag New York, Inc.; New York.* GREINER, WALTER; BROMLEY, D. ALLAN (FOREWORD).

9 Classical Electrodynamics **1998:** *Perseus books; Reading, Massachusetts.* SCHWINGER, JULIAN; DERAAD, LESTER L. JR.; MILTON, KIMBALL A.; TSAI, WU-YANG.

10 Advanced Classical Electromagnetism **2022:** *Princeton University Press; Princeton.* WALD, ROBERT M.

11 Fundamentals: Ten Keys to Reality **2021:** *Penguin Press, New York.* WILCZEK, FRANK.

12 Newton on Action at a Distance **2014:** *J. Hist. Phil. 52(4), 675.* DUCHEYNE, STEFFEN.

13 The Variational Principles of Mechanics **1952:** *Oxford University Press; London.* LANCZOS, CORNELIUS.

14 Advanced Classical Electrodynamics: Green Functions, Regularizations, Multipole Decompositions **2017:** *World Scientific Publishing Co. Ptd. Lte.; Singapore.* JENTSCHURA, ULRICH D.

15 There are no particles, there are only fields **2013:** *Am. J. Phys. 81(3), 211.* HOBSON, ART.

16 Classical Electrodynamics in Terms of Direct Interparticle Action **1945:** *Rev. Mod. Phys. 21(3), 425.* WHEELER J. A.; FEYNMAN R. P.

17 Relativistic Hamiltonian Dynamics in Nuclear and Particle Physics, **1991:** *Adv. Nucl. Phys. 20, 225.* KEISTER, B. D.; POLYZOU, W. N.

18 Action at a Distance and Cosmology: A Historical Perspective **2003:** *Annu. Rev. Astron. Astrophys, 41(1), 169*. NARLIKAR, J. V.

19 Introduction to Elementary Particles; Second, Revised Edition **2008:** *WILEY-VCH Verlag GmbH & Co.; Weinheim*. GRIFFITHS, DAVID.

20 The Rest-Frame Darwin Potential from the Lienard-Wiechert Solution in the Radiation Gauge **2001:** *Ann. Phys. (N. Y.) 289(2) 87*. CRATER, HORACE; LUSANNA, LUCA.

21 What is the Force Between Electrons? **1998:** *Adv. Quantum Chem. 30, 433*. SUCHER, J.

22 Alternative routes to the retarded potentials **2017:** *Eur. J. Phys. 38(5), 055203*. HERAS, RICARDO.

23 The present status of Maxwell's displacement current **1998:** *Eur. J. Phys. 19, 155*. ROCHE, JOHN.

24 On Maxwell's displacement current **1998:** *Eur. J. Phys. 19, 469*. JEFIMENKO, OLEG D.

25 Reply to Comment 'On Maxwell's displacement current' **1999:** *Eur. J. Phys. 20, L21*. ROCHE, JOHN.

26 Maxwell's displacement current revisited **1999:** *Eur. J. Phys. 20, 495*. JACKSON, J. D.

27 Reply to J D Jackson's 'Maxwell's displacement current revisited' **2000:** *Eur. J. Phys. 21, L27*. ROCHE, J.

28 Reply to Comment by J Roche on 'Maxwell's displacement current revisited' **2000:** *Eur. J. Phys. 21, L29*. JACKSON, J. D.

29 A formal interpretation of the displacement current and the instantaneous formulation of Maxwell's equations **2011:** *Am. J. Phys. 79(4), 409*. HERAS, JOSÉ A.

30 Personal communication.

31 Instantaneous Action-at-a-Distance in Classical Relativistic Mechanics **1967:** *J. Math. Phys. 8(2), 201*. HILL, ROBERT NYDEN.

32 Instantaneous action-at-a-distance representation of field theories **1993:** *Phys. Rev. E 48(5), 4008*. VILLECO, R. A.

33 Abraham-Lorentz versus Landau-Lifshitz **2010:** *Am. J. Phys. 78(4), 391*. GRIFFITHS, DAVID J.; PROCTOR, THOMAS C.; SCHROETER, DARRELL F.

34 An introduction to the Lorentz-Dirac equation **1999:** *arXiv:gr-qc/9912045*. POISSON, ERIC.

35 Asymptotics of a proposed delay-differential equation of motion for charged particles **2005:** *arXiv:gr-qc/0205065v8*. PARROTT, STEPHEN.

36 The Motion of Point Particles in Curved Spacetime **2004:** *Living Rev. Relativ. 7, 6*. POISSON, E.

37 Analytical Mechanics **1998:** *Cambridge University Press; New York*. HAND, LOUIS N.; FINCH, JANET D.

38 Classical Mechanics; Third Edition **2001:** *Addison-Wesley, San Francisco*. GOLDSTEIN, HERBERT; POOLE, CHARLES P. JR.; SAFKO, JOHN L.

39 Classical Mechanics Systems of Particles and Hamiltonian Dynamics (Classical theoretical physics) **2003:** *Springer-Verlag New York, Inc.; New York*. GREINER, WALTER; BROMLEY, D. ALLAN (FOREWORD).

40 Classical Mechanics, Point Particles and Relativity (Classical theoretical physics) **2004:** *Springer-Verlag New York, Inc., NY*. GREINER, WALTER; BROMLEY, D. ALLAN (FOREWORD).

41 Analytical Mechanics for Relativity and Quantum Mechanics **2005:** *Oxford University Press Inc.; New York*. JOHNS, OLIVER DAVIS.

42 A new kind of cyclic universe **2019:** *Phys. Lett. B 795, 666*. IJJAS, ANNA; STEINHARDT, PAUL J.

43 Classical Relativistic Many-Body Dynamics **1999:** *Springer Science+ Business Media; Dordrecht*. TRUMP, M. A.; SCHIEVE, W. C.

44 Mathematical Foundations of Quantum Theory **1978:** *In Mathematical Foundations of Quantum Theory; Marlow, A. R. (Ed); Academic Press, Inc.; New York*. DIRAC, P. A. M.

45 Action-at-a-distance as a full-value solution of Maxwell equations: The basis and application of the separated-potentials method **1996:** *Phys. Rev. E 53(5), 5373–5381*. CHUBYKALO, ANDREW E.; SMIRNOV-RUEDA, ROMAN.

46 Erratum: Action-at-a-distance as a full-value solution of Maxwell equations: The basis and application of the separated-potentials method [Phys. Rev. E 53, 5373 (1996)] **1997:** *Phys. Rev. E 55(3), 3793—3793*. CHUBYKALO, ANDREW E.; SMIRNOV-RUEDA, ROMAN.

47 Comment on "Action-at-a-distance as a full-value solution of Maxwell equations: The basis and application of the separated-potentials method" **1998:** *Phys. Rev. E 57(3), 3680—3682*. IVEZIĆ, TOMISLAV; ŠKOVRLJ, LJUBA.

48 Reply to "Comment on 'Action-at-a-distance as a full-value solution of Maxwell equations: The basis and application of the separated-potentials method' " **1998:** *Phys. Rev. E 57(3), 3683—3686*. CHUBYKALO, ANDREW E.; SMIRNOV-RUEDA, ROMAN.

49 Exponential integrators for stochastic Maxwell's equations driven by Itô noise **2020:** *J. Comput. Phys. 410, 109382*. COHEN, DAVID; CUI, JIANBO; HONG, JIALIN; SUN, LIYING.

50 Multi-symplectic discontinuous Galerkin methods for the stochastic Maxwell equations with additive noise **2022:** *J. Comput. Phys. 461, 111199*. SUN, JIAWEI; SHU, CHI-WANG; XING, YULONG.

51 Stochastic Lorentz forces on a point charge moving near the conducting plate **2008:** *Phys. Rev. D 77, 105021*. HSIANG, JEN-TSUNG; WU, TAI-HUNG; LEE, DA-SHIN.

52 Fluctuating, Lorentz-force-like coupling of Langevin equations and heat flux rectification **2017:** *Phys. Rev. E 96, 022109*. SABASS, B.

53 Fokker-Planck analysis of the Langevin-Lorentz equation: Application to ligand-receptor binding under electromagnetic exposure **1997:** *J. Appl. Phys. 82(9), 4669*. MOGGIA, ELSA; CHIABRERA, ALESSANDRO; BIANCO, BRUNO.

54 Covariant Electrodynamics; A Concise Guide **2005:** *The Johns Hopkins University Press; Baltimore*. CHARAP, JOHN M.

55 Electrodynamics and Classical Theory of Fields & Particles **1980:** *Dover publications, Inc; New York*. BARUT, A. O.

56 Symmetries and conservation laws: Consequences of Noether's theorem **2004:** *Am. J. Phys. 72(4), 428*. HANCA, JOZEF; TULEJAB, SLAVOMIR; HANCOVAC, MARTINA.

57 Energy conservation in explicit solutions as a simple illustration of Noether's theorem **2019:** *Am. J. Phys. 87(2), 141*. PÖSSEL, MARKUS.

58 Advanced general relativity **1996:** *Cambridge University Press; New York*. STEWART, JOHN.

59 A first course in general relativity **2004:** *Cambridge University Press; Cambridge*. SCHUTZ, BERNARD F.

60 General relativity; second edition **2013:** *Springer Science+Business Media Dordrecht; New York*. STRAUMANN, NORBERT.

61 General Relativity **1984:** *The University of Chicago Press; Chicago*. WALD, ROBERT M.

62 Survey of Fundamental Laws **1971:** *In Physical Chemistry: An Advanced Treatise; Vol 1, Thermodynamics; Wilhelm, Jost (Ed); Academic Press Inc.; New York*. HAASE, R.

63 Fundamental Physical Problems of Quantum Electrodynamics **1980:** *In Foundations of Radiation Theory and Quantum Electrodynamics; Barut, A. O. (Ed); Springer Science+Business Media; New York*. ROHRLICH, F.

64 Quantum Electrodynamics (Advanced Book Classics) **1998:** *Westview Press*. FEYNMAN, RICHARD P.

65 Quantum field theory; 2nd edition **2010:** *John Wiley & Sons, Ltd, Chippenham*. MANDL, FRANZ; SHAW, GRAHAM.

66 Genius: The Life and Science of Richard Feynman **1992:** *Pantheon Books; New York*. GLEICK, JAMES.

67 Cosmology and action-at-a-distance electrodynamics **1995:** *Rev. Mod. Phys. 67(1), 113*. HOYLE, F.; NARLIKAR, J. V.

68 Casimir effect and the quantum vacuum **2005:** *Phys. Rev. D 72, 021301*. JAFFE, R. L.

69 Proof that Casimir force does not originate from vacuum energy **2016:** *Physics Letters B 761, 197*. NIKOLIĆ, HRVOJE.

70 The Notions of Localizability and Space: From Eugene Wigner to Alain Connes **1989:** *Nucl. Phys. B Proc. Suppl. 6, 222*. BACRY, HENRI.

71 Magnetism of matter and phase-space energy of charged particle systems **1999:** *J. Phys. A: Math. Gen. 32, 2297*. ESSÉN, HANNO.

POSTFACE

Thank you for reading! I hope that you enjoyed this book. And now what? The first step is to share the advanced knowledge you have acquired. Please consider leaving an honest review on your favorite online store or on social media, or consider recommending this book to friends, colleagues, and family. The sooner we remove these myths from popular and academic literature, the better it will be for everyone. The second step is to address the inconsistencies and incompatibilities of current formalisms and theories. I hope I have motivated my colleagues to review physics from top to bottom.

What is the best way to improve the formalisms and current theories is open to debate. I have dubbed ***symplectics*** my own approach. My ultimate goal is to provide *a unified and self-consistent formulation for physical, chemical, and biological processes*. There is only one universe, the Natural World, and it seems evident that a single theory should be able to describe it.

Made in the USA
Monee, IL
13 May 2025

d3adb304-6a6c-4abe-a7fa-b2d0a57d610aR01